Assessment of Improved Crop Production Technologies

Assessment of Improved Crop Production Technologies

Editor
Dr. Ajoy Gangopadhyay

2007
Gene-Tech Books
New Delhi - 110 002

ISBN 81-89729-63-2

Published by : **GENE-TECH BOOKS**
4762-63/23, Ansari Road, Darya Ganj,
NEW DELHI - 110 002
Phone: 41562849
e-mail: genetechbooks@yahoo.co.in

Printed at : **Tarun Offset Printers**
Delhi

PRINTED IN INDIA

Preface

Low agricultural productivity in developing countries is a common feature. One of the reasons is non-adoption of improved and recommended technologies. Crop production technologies are there but their adoption is low. Increase in the productivity has always remained a concern of scientists. Many technologies have developed by the researchers to improve the efficiency of crop production systems. In agriculture, production technology may be an input, use of an implement, practice or a procedure to overcome a difficulty, to increase yield or save time.

By adopting improved crop production technologies, farmers can get yields closer to potential yields. However, the awareness about exotic inputs and non-traditional methods of seed treatment, pest and disease control is comparatively low. The knowledge does not guarantee the adoption of technologies. It depends on the flow of information system, farmers' resources, behaviour and the socio-economic conditions. Many crop production technologies, those based on indigenous knowledge, have widely been adopted. However, super impositions of non-traditional treatments on local technologies have made the adoption partial. These include chemical treatments like seed treatment and chemical weed control and use of pesticides. To enhance the adoption of technologies, which ultimately leads in reducing the gap between actual and potential yields creation of awareness is very important. The strengthening of linkages between research, extension and farmer is also essential.

This book assesses the adaptability of improved crop production technologies by farmer community. In a writing style that is clear, to-the-point, and easy to understand, this book provides the necessary background for a solid understanding of all issues related to the topic. The book will be suitable for use in courses covering crop production at either the introductory or intermediate levels of study and will be an invaluable reference for students and professionals alike.

Editor

Contents

1

Improved Crop Production

The rapid population growth in general and the fact that most of the population growth takes place in developing countries rises the question how these additional people can be fed. Basically, there are two different possibilities to increase food production: Enlarge crop yields on already cultured land, or expand the area of cultured land.

The production of rice, wheat and maize, which are the most important crops for human nutrition, doubled during the last 25 years. Higher yields were possible because of the introduction of high-yielding varieties, improved cultivation techniques, fertiliser use and chemical crop protection measures - a development which has been termed "green revolution".

The intensification of crop production improved the level of food consumption in most developing countries during the last 30 years despite a world population growth of 2.4 billion in the same period. However, the intensification also led to a number of disadvantageous side-effects like erosion, deterioration of soil quality, salinisation, and others.

Every year, about 1 % of the cultured area has to be replaced because of the above mentioned reasons.

Especially in developed countries, the use of agrochemicals is a growing public concern. The amount of crop protection products used has risen steadily during the past 30 years. Presently, more than 25 billion US$ per year are spent for herbicides, fungicides, insecticides, soil disinfectants and growth regulators. In Western Europe, the most crop protection products are used per area of arable land, followed by the USA.

Plant genetic engineering may contribute substantially to the development of new pest and disease resistant varieties by increasing the speed of conventional breeding and by transferring useful genes from various plants and other organisms into crops.

In Asia, most vegetable growers have a land holding which is smaller than one hectare (2.4 acres). They are interested in any technology which helps them to produce more crops each year from their land, particularly during the off season when prices are higher. For this reason, greenhouses and other types of structures have become popular among Asian growers. Production under structures not only increases the total annual yield per unit area, but also improves the quality. Vegetables have a better flavor and appearance when they are grown in a controlled environment, where growers can make precise adjustments of nutrients and other inputs.

The kinds of protective structures used by growers in Asia range from simple structures such as rain shelters, shadehouses, mulches, row covers and plastic tunnels, to permanent structures covered in plastic or glass with computerised environmental controls. Growers in countries where typhoons are common over the summer tend to prefer low-cost structures which can be quickly and cheaply replaced. Greenhouses in such countries generally have walls and roof of plastic rather than glass, over a metal or even a bamboo framework.

The discovery and development of the polyethylene polymer in the late 1930s, and its subsequent introduction in the early 1950s in the form of plastic films, mulches, and drip-irrigation tubing and tape, revolutionised the commercial production of many vegetable crops. The later discovery of other polymers, such as polyvinyl chloride, polypropylene, and polyesters, and their use in pipes, fertigation equipment, filters, fittings and connectors, and row covers, further extended the use of plastic in agriculture.

In general, structures are used to overcome low temperatures in temperate Asian countries, and high temperatures in tropical ones. Modifying the temperature is required, not only for crop protection, but also to provide growers with more comfortable working conditions. A common use of structures in both cool and hot climates is for the nursery production of high-quality and/or early seedlings for transplanting.

In Japan, greenhouses were first introduced to protect crops from cold weather. Now greenhouses are commonly used all year round. High summer temperatures are modified by good ventilation, shading and evaporative cooling. New covering materials are being developed which block solar radiation. In Malaysia, the area used for the protected cultivation of horticultural crops has expanded rapidly over the last decade.

The most important constraint is marketing. In general, vegetables produced under structures cannot successfully compete with those produced in the open field imported from neighboring countries, unless stringent measures are taken by the authorities to safeguard the quality and safety of vegetables. These include specifying the maximum levels of pesticide residues, and accurate labeling.

Greenhouse production of vegetables in Korea has made much progress in terms of the external design and the construction of modernised glasshouses. However, there is still a need to reduce the labor cost of operations such as transplanting and harvesting. Protected cultivation of vegetables takes 2-3 times more labor than open field cultivation. Mechanisation of some operations may be necessary. This in turn implies standard facilities where machines can be used.

In Taiwan, some fruits such as peaches are being produced under structures. Establishing a permanent orchard structure is expensive, and needs a considerable amount of technical skill. The system must be able to meet the environmental requirements of the fruit grown, so that trees produce a stable yield of high-quality produce.

In Thailand, protective structures are being used to produce flowers by the hill tribes of northern Thailand, under the Royal Project Foundation. At present, the Royal Project Foundation has a large planting area under plastic in Thailand, 90% of which is used for flower production. Greenhouse designs developed by the Foundation have already been widely used nationwide. Structures for some fruit crops such as grape and peach are under investigation and show promising results, while research on vegetables under structures is also under way.

While greenhouse production increases the net income of farmers in Northern Thailand, they still need a great deal of outside technical support. Greenhouse production is so different from their traditional agriculture, that it will take some time for growers to acquire all the needed technical skills. This is a common problem elsewhere in Asia. Production under structures requires good management. Proper planning and attention to detail are needed, if the grower is to succeed

Another widespread problem is that small-scale farmers may find it difficult to finance the investment cost of even a simple structure. In the Philippines, for example, the initial cost of establishing protective structures such as greenhouses and plastic houses is considerable, and not within the reach of ordinary farmers. Designs using cheaper materials are needed. Furthermore, mulching materials which are durable and biodegradable must be developed to protect the environment. Further studies are also needed on protective structures which can withstand typhoons, and which have a better ventilation system for a tropical climate.

Mechanised Production

Whereas many Asian smallholders have adopted machinery for rice production, machines are not much used under structures. There are several reasons for this. Firstly, most farms are too small. Even if suitable machines were available, the utilisation rate would be very low. Secondly, growers produce a wide range of crops. It is not easy to develop machines that can take care of a wide range of needs of a variety of crops. Thirdly, most greenhouses are temporary structures with narrow walkways which are difficult for machines to enter.

However, the scale of farming in most Asian countries is expected to increase, particularly in highly industrialised ones. Many growers are enlarging the size of their operations through group farming or producers' associations. Designs for greenhouse machinery should focus on attachments which can be adapted to various crops and different operations, and simple, lightweight machinery suitable for a rural labor force which includes many women and old people. Eventually, new cultivation models with uniform beds and standardised seedling trays can be expected for mechanised operations.

Vegetable Seedlings

In vegetable production in greenhouses, continuous cultivation leads to heavy losses, particularly from soil-borne diseases and nematodes. Grafted seedlings have become popular in Japan. Grafting gives greater disease tolerance and increased vigor to crops, to an extent which justifies the higher cost of such seedlings. In 1990, in areas producing watermelons, cucumbers, melons, tomatoes and eggplants, 59% of plants were grafted. A range of rootstock species and varieties are used, and various grafting methods. Recently, tube grafting has become popular in the manual grafting of tomato, eggplant and cucumber plants, because it is faster. Grafting robots and healing chambers have been developed for use in plant nurseries.

Plug Culture System

In plug culture, vegetables are sown by direct seeding into small cells or plugs. Usually a single seed is planted in each plug. When the seedlings have grown to the 2-3 leaf stage, they are transplanted into flats. These may contain from 18 to more than a hundred cells full of growing medium. The flats are usually left to grow under netting or some kind of structure. The plug system is widely used in Taiwan for the production of leafy vegetables, which are harvested at an early stage when they are still young and tender.

Consumers like the flavor of the vegetables, and the fact that they are free of field soil. From the grower's point of view, plug seedlings have a faster growth rate, a shorter cropping period and a longer storage life. Research data also shows that vegetables grown in a porous medium have better root and shoot growth than those grown in field soil, resulting in higher yields. However, the system is labor intensive, particularly transplanting the plugs into the flats and taking them to the field. Future research should focus on mechanisation and automation of the system to reduce the labor demand. Another useful innovation would be biodegradable or photodegradable flats that are more environmentally friendly. The possibility of using colored flats which reflect light, thus promoting plant growth and repelling insects, should also be explored.

Plug production technology was introduced into Korea in 1992, and is now an important industry. The technology was first used for the mass production of vegetable seedlings to be transplanted into greenhouses and the open field. This gave growers a steady supply of high-quality seedlings, and relieved them of the burden of having to produce their own planting materials. Recently, there has been growing interest in flower plug transplants.

Soilless Culture

Since the early 1980s, many soilless growing systems have been developed in Asia, some suited to tropical countries and some to temperate ones. Some of these were based on hydroponics. Early types were open systems. More recent ones are closed. Drained water is collected and brought to a central poinmt for reuse, after the water has been disinfected and its EC and pH adjusted. Nowadays, however, 80% of greenhouse producers are using some kind of artificial medium for their crops. Soilless systems using media of various kinds systems give growers good control over plant nutrients, so that produce has good appearance, flavor and nutritive value. Porous media also offer an ideal environment for root growth.

Agricutlural Technology for Value-added Food Production

Agricultural production is a prerequisite to rural poverty alleviation. However, in itself, agricultural growth is not sufficient to eradicate poverty, nor can it create adequate opportunities for enhanced rural employment, income generation, and economic access to food.

Agro-processing maintains a high potential for boosting economies based on agriculture to help ensure sustainable growth and increasing rural incomes in the region. And while the new global economy ushers in a rapid expansion of the worldwide distribution of agricultural products, our concern now is not only to produce food in sufficient quantity, but also to enhance the competitiveness of the Asian agricultural industry in the world market by developing value-added foods based on higher quality standards and improved processing techniques.

Scientific and Technological Development

In the last half of the 20th century, rapid industrialisation and urbanisation have been consuming too much natural energy and resulting in rapid migration of humans and the fast movement of resources and information. Agriculture is not exempted from this fast change and it is now shifting from small-scale to commercial and even to global production. Thus, new agricultural technologies have been developed in order to supply stable and sufficient food to the expanding population. Meanwhile, sustainable agriculture as well as the preservation of the diverse food culture and maximum utilisation of food and biomass must be ensured. East Asia is the origin of diverse agricultural technology and food culture, and is envisioned to be at the forefront of scientific and technological advancement in agricultural production.

The extensive application of diverse agricultural technologies is vital in order to meet the demands of an increasing population. Agricultural production in the Asian region is notably decreasing, with the rate of increase in food production less than that of the increase in population, particularly in China and India. Unless changes in agricultural policies are made, the threat of hunger and famine will continue to loom in some developing countries in the region.

Efficiency of Land use

More efficient land use is necessary in view of the increasing demand in agricultural products and the decreasing rate in annual production. Optimum production rate must be met in order to keep the sustainable use of land, i.e. crop yield and fertiliser use must be balanced.

Improved Crop Varieties

Increased crop yield must also be achieved through the use of better varieties with higher photosynthetic activities, and better utilisation of phosphorus, potassium, and other minerals under arid conditions such as low or high pH soil, high salinity, and undesirable water conditions. These varieties must be developed as soon as possible in order to cope with the rapid increase in population and erosion of fertile soils through new biotechnology such as recombinant technology and land management.

Raw materials Production

An understanding of agroecosystems is the key to determining effective farming systems. A report of results from a 21-year study of agronomic and ecological performance of biodynamic, bioorganic, and conventional farming systems in Central Europe showed crop yields to be 20% lower in the organic systems, although input of fertiliser and energy was reduced by 34% to 53% and pesticide input by 97%. Enhanced soil fertility and higher biodiversity found in organic plots may render these systems less dependent on external inputs.

Improvement of yields of crops can also be achieved by optimising traditional or conventional farming, organic farming, and bio-farming or the use of genetically modified crops. These three farming types can co-exist depending on the farming and land conditions. They can complement each other and actually are not competitive, for each will have its own market. It is obvious however that organic farming will cost much in terms of label and management so that the selling price is naturally higher. Traditional or conventional farming may sacrifice sustainability of land use. Thus, bio-farming and good management is the one expected to give the least cost and best sustainability. However, recombinant crops must first be proven to be safe to humans, animal and the environment.

Value-added Food Processing

Traditional foods are a treasure box of resources in terms of material and knowledge. The way they are produced, the local substrates used in the countries where they are developed, and the presence of diverse microorganisms in these foods are good and novel resources for the development of technology for the food industry. These microbes and fermented products have various functionalities in maintaining human health, thus, have great potential in the development of physiologically functional foods. Oligosaccharides with mineral enhancing capability, prebiotic activity, anti-constipation, peptides with anti-hypertensive, improvements of lipid metabolisms, anti-obesity, and cholesterol-lowering activity are found in various Asian traditional foods.

Some Asian traditional foods with potential physiological functions and which should be given emphasis for quality improvement through research and development (R&D) include: brem (rice wine), ragi (rice paste), tempeh (fermented soybean), kimchi (fermented cabbage), onchom (fermented soybean), urutan (fermented sausage), nampla (fermented fish sauce), plarah (fermented fish), saigokisan (fermented sausage), balao-balao (fermented rice and shrimp), burong babi (fermented pork), burong kanin (fermented rice and fish), longanisa (fermented sausage), patis (fermented fish sauce), tapa (fermented beef), toyo, (soy sauce), bagoong (fish paste), nata de coco, natto, bonito, yogurt, and calpis.

Biotechnology in Production of Seeds and Seedlings

Over the past two decades, the technology for producing seeds and seedlings has developed very rapidly. Many new techniques have been incorporated into production systems, and have benefited farmers all over the world. However, this rapid pace of technical development means that scientists and technicians involved in the production of high-quality seeds are constantly having to update their skills.

Genetic diversity is the basis of plant breeding. Precious biodiversity is being lost with the disappearance of numerous wild plant species, mainly because of overexploitation and loss of habitat. Many wild plants have medicinal uses. They tend to become more valuable the scarcer they become. This encourages people to gather them to sell, and puts heavy pressure on surviving populations.

Tissue Culture Techniques

Tissue culture techniques are being widely used for the propagation and conservation of many traditional Chinese medical plants. Some of them are becoming rare in the wild. The conservation of germplasm *in vitro* keeps these species safe from accidental extinction. The techniques also make available a large body of material for pharmacological studies and chemical analysis. They also provide large quantities of disease-free seedlings for farmers. Tissue culture has made possible the mass production of disease-free and uniform plants. The technique thus brings farmers the great benefit of high-quality planting materials of new high-value crops

Finally, these same techniques can be used for the safe transfer of germplasm from one country to another. Many of the diseases and pests which damage agriculture today were introduced accidentally when new crop varieties were imported from overseas. So much damage has already been done from these accidental invasions, that it is becoming difficult for countries to exchange plant materials. Many governments tend to feel that the possible damage from introductions may well outweigh any potential benefits. However, plant breeding programs need access to new germplasm. Tissue culture offers a solution to the problem. Germplasm can cross international boundaries safely, in the assurance that no pests and diseases are secretly traveling with it.

Techniques for Disease Detection

In recent years, remarkable progress in molecular biology

techniques has enabled scientists to acquire a more detailed understanding of plant pathogens. This training course introduced participants to current techniques of serological and molecular diagnostic techniques.

Serological assay is based on antibodies produced by animals, usually rabbit or rats. When suitable antibodies are available, the most sensitive and efficient test for plant virus is ELISA (enzyme-linked immunosorbent assay). ELISA is widely used, since it is easy to apply and does not need any expensive equipment.

Another testing method is PCR (polymerase chain reaction). PCR testing is very accurate, but it is also rather expensive and technically difficult. The participants were shown the preparation of samples for PCR testing, including the precautions needed to avoid the contamination of samples. They then practiced the steps themselves in laboratory exercises.

DNA Fingerprinting

The DNA composition of an individual plant is what determines its essential characteristics. The DNA fingerprint describes the DNA sequence profile of an individual. This is the same in every cell of that individual. The DNA fingerprint cannot be altered by any change in the environment. Nor can it be changed by anything that happens to an individual during its lifetime.

DNA markers

Different DNA markers are available, including RFLP (restriction fragment length polymorphism), RAPD (random amplified polymorphic DNA) and AFLP (amplified fragment length polymorphism). These markers are used to characterise a variety or line, and in this way protect the intellectual property rights of the plant breeder. They are also used to evaluate the purity of seeds, and to detect genetic drift in germplasm collections. It is also advisable to use the material for only five generations, or only five sub-culturing cycles.

Future Challenges

To meet the food requirements of the increasing Asian population in the next decades, agricultural food production must be increased, not just in terms of quantity but also in quality. The development and promotion of high-quality and value-added agro-products are vital in response to the changing market and consumer preferences amid recent trends in economic development and globalisation, as well as the expanding population.

Asian countries must realise the limitation of agricultural productivity in terms of human and natural resources. Hence, increased agricultural food production must be based on technological improvement to promote: a) sustainable, efficient and economic use of land and other resources; b) crop variety improvement; c) value-added approaches in food processing to develop various functionalities in response to changing market and consumer preferences; and d) development of quality standard for processed food for global competitiveness.

To ensure sustainable economic and agricultural development in Asia, food production must be viewed in terms of rural poverty alleviation. Agro-processing must be able to create adequate opportunities for enhanced rural employment, income generation, and economic access to food. Strategic alliance or integration between small-scale food producers and the manufacturing industry must be enhanced, possibly through farmers' organisations.

With the diverse traditional food culture in each country in Asia, there is a great potential for the development of rich and high-quality food resources within the region. We must recognise and re-evaluate the value and importance of the Asian food culture in response to changing consumer preference, the demand for more food variety, and the promotion of physiologically functional foods. However, we must study traditional food processes in detail and learn from these conventional technologies, as well as develop appropriate new technologies to achieve high-quality standard for traditional food.

2

Crop-Livestock Technologies

Most livestock products in mixed farming systems are derived from animals that are fed on local resources such as pasture, crop residues, fodder trees and shrubs. Another feed resource consists of by-products from village industries such as oil extraction, grain milling and cane crushing. The exchange of these resources saves money and reduces waste by recycling products within the farming system. They create employment and they can contribute to soil texture and fertility while being an economic incentive for rearing multipurpose animals.

For example, in expansion agriculture (EXPAGR), crops and soils are not strongly linked. Animals tend to graze elsewhere, they are a savings account and they may bring home some dung and young stock to pull the plough. However, in new conservation agriculture (NCA) the crops and livestock are integrated, as in Low external input agriculture (LEIA), and they both serve the same goals - to keep the farming sustainable and productive. Still, under poverty conditions in LEIA, there is a tendency for good-quality feed residues to be sold off the farm, thus leaving the animals with mainly fibrous feeds. Therefore animals in LEIA tend to be a way to convert fibrous feeds, crop residues, poor-quality grazing land

and weeds into draught, dung and urine, often at the expense of the land in High external input agriculture (HEIA) there is hardly any attention to interactions among soil, plants and animals because limiting nutrients can be bought from outside and waste is ignored until it becomes a serious problem. HEIA exists not only in temperate and rich countries, but also in tropical and poor countries.

Crop Residue and Soil Management

Livestock can enhance farm output by intensifying nutrient and energy cycles, particularly in LEIA and NCA. In LEIA this mainly implies a closer recycling of nutrients in straws, dung and urine. In NCA there is scope to include leys, anti- erosion bunds and, for example, adjusted cropping patterns. NCA also has the added advantage that it can both import some nutrients in fertiliser and feed while it aims to keep the better-quality crop residues such as brans and cakes on the farm. Crop residues can be used for animal feed as well as for soil fertility and can also be sold. The farmers have to choose how to use these materials and, whatever they are used for, the choice needs to be accompanied with keen management to make the best of these resources. The fibrous crop residues have a potential in soil conservation through mulching but even the oilseed cakes can be used to improve soils. Some terms related to animal-plant-soil interaction are explained in Table 1.

Cycling of Nutrients

The use of dung and urine excreted by cattle, pigs, chickens and other animals for crops is traditionally raised as an argument in favour of keeping livestock. Dung and urine contain several nutrients such as nitrogen, phosphorus and potassium, and the solid fraction contains organic matter that is important to maintain soil structure and fertility. Occasionally it is even believed that dung and urine have medicinal qualities, but more commonly the dung is used as fuel, while it can also be used to plaster walls when mixed with clay. The belief in the medical value may have some justification, e.g. through the occurrence of antibiotics in dung and through the presence of lignin derivatives in dung that help to make it sticky or even to give it bactericidal properties.

Table 1. Terms and technologies associated with crop-livestock integration, soil fertility and animal nutrition

Fallow	Allowing the land to remain idle so that it can rebuild itself in terms of soil fertility, soil structure and soil life; or to ensure that disease and weed pressure will be reduced
Ley	A cultivated fallow planted with fodder crops that actively suppress weeds or disease, bind nitrogen, mobilise phosphor us, add soil organic matter or reduce runoff, etc.
Infield	The cropping area around villages or homesteads that is fertilised with excreta and with litter from the outfield
Outfield	The grazing lands, waste lands and forests around the village where animals are grazed or the faraway crop fields of a village
Catch crop	A crop that is planted to "absorb" nutrients in the soil that are released through weathering
Anti-erosion bund	A ridge planted (on the edge of a terrace) to avoid runoff
Straw	The stems and leaves of mature fine cereals such as rice, wheat, barley, oats and rye
Stover	The mature stems and leaves of coarse cereals such as maize, sorghums and millets
Grain:straw ratio	The ratio between grain and straw, i.e. a grain yield of 2 000 kg/ha associated with a grain:straw ratio of 1.5 results in a straw yield of 3 000 kg/ha

Contd....

Contd....

Harvest index	The proportion of the above ground biomass (in the case of cereals) that is found in the grain, i.e. a harvest index of 33 percent results in a ratio of 2 kg straw:1 kg grain
Excreta	A combination of dung and urine
Manure	Farmyard manure, excreta (the solid part) of animals that is to a lesser or greater degree mixed with straws and other leftovers
Soil organic matter	Organic matter originating from plants and/or dung that occurs in several degrees of decomposition in the soil, and in different degrees of solubility. It serves to enhance structure (on heavy soils) and water/nutrient holding capacity (on light soils)
Zero grazing	The keeping of animals in a shed, based on feed that is brought in from elsewhere (on-farm and/or off-farm)
Mulch	The mix of dry and/or green materials that can be used to cover the soil for prevention, for example, of erosion or excessive temperature effects
Compost	Organic matter (with or without animal excreta) that is decomposed on a heap or in a pit before applying it to the soil

The use of dung and urine for soil fertility is most cited. It is particularly important in LEIA and NCA, but unimportant in EXPAGR. The term dung "production" is misleading because animals excrete only nutrients that they first have to eat. Animals can carry nutrients from one place to another, they can increase the rate of their turnover or they can concentrate the nutrients from the outfield to the infield, i.e. they concentrate the fertility of a large land area on relatively small cropping areas around the villages. The "production" of dung in HEIA is a disadvantage rather than a blessing and elaborate systems are designed to dispose of dung rather than to use it.

Organic matter comes from manure, household waste and compost, or from incorporating weeds into the soil.

The tasks involved in producing and using organic fertiliser are divided among different members of the household. Young people or labourers tend to collect the crop residues, to dig compost pits when needed and to transport the organic fertiliser to the field while the head of the family is responsible for managing its production. Women make organic fertiliser from household waste and keep some of it back for use on their own small plots of land. The major nutrients in fertiliser and dung are needed to make plants grow and produce:

— Nitrogen is essential for growth, for making the stems, leaves and roots; it is an important component of proteins.

— Phosphorus is needed for the plants to flower, to bear fruits and to develop strong roots.

— Potassium is important for tuber and fruit enlargement while it helps to maintain the healthy activity of all plant tissues.

Cycling of Nutrient Manure

The amount and quality of urine and dung produced depends on the type of animal, its size and the type of feed as well as on the management of the farmer. One way to calculate the amount of manure produced is:

— One animal of 250 kg live weight has a feed intake that averages 2.5 percent dry matter of its live weight. It therefore consumes 250*365*0.025 = 2 280 kg of dry matter. With an average digestibility of 55 percent, the animal will produce 0.45*2 280 = 1 026 kg of dung every year.

— Small ruminants weighing 25 kg and feeding 3.2 per cent daily on average of their live weight, consume 25*365*0.032 = 292 kg dry matter. If their average digestibility of the feed is estimated at 60 percent it can be calculated that one small ruminant produces some 117 kg of dry matter faeces per year.

After the harvest a part of the crop residues tends to be freely accessible to all livestock including those from outside the village. The residues consist of straw that is sometimes left in the field or carried home and they contain stubble and ratoon which can be grazed. Another part of the residue is fallen grain - in Asia this is used by ducks. Weeds can also be considered as crop residues and in wetter areas or seasons they can form an important feed resource. Sometimes they are cut and carried home during the cultivation period when access to grazing grounds is difficult for animals.

The nutrient content of manure and other organic fertilisers varies according to the quality of feed and the way it is stored and handled. Table 2 gives an indication of the concentration of the main nutrients in the manure of cattle and small ruminants and other sources of organic fertiliser. Dry matter content of manure also varies widely; in cows on lush pasture it can be less than 15 percent but in sheep on dry forage it can be higher than 50 percent. The amount and proportion of nitrogen excreted depend on animal diet. The urine and solid dung of animals fed highly digestible diets with a lot of protein contain much more nitrogen and, therefore, are more susceptible to nitrogen losses than excreta from diets containing greater amounts of roughage. Much of the urine nitrogen is lost via ammonia volatilisation.

Where animal management tends towards increased stall-feeding, the composting of fresh manure will have to play a greater role in minimising nutrient losses. Pits or heaps that capture feed refusals, manure and urine and household waste need to be designed to minimise nutrient losses. Low-cost appropriate implements to spread the compost at the appropriate time over large cultivated areas such as the Sahel are also needed.

Table 2. Nutrient contents of manure and other organic fertilisers (percentage)

Organic fertiliser	*N*	*P*	*K*	*Dry matter*
Cattle				
Fresh manure	1.4-2.8	0.5-1.01	0.5-0.6	15-25
Kraal (litter)	0.5-2.3	0.22-0.81	0.77-5.44	40-60
Kraal (no litter)	1.5-2.5	0.2-0.6	1.5-2.0	30-50
Goat and sheep				
Fresh manure	2.2-3.7	0.25-1.87	0.88-1.25	50-70
Fresh green manure	2.0-4.3	0.1-0.3	1.0-3.4	
Compost	0.3-0.9	0.07-0.17	0.14-1.3	
Household waste	0.2-0.9	0.05-0.5	0.1-2.1	
Ash from farm cooking	0.2-0.6	0.1-0.6	1.1-2.7	

In various systems draught oxen, donkeys, horses and small ruminants are kept overnight in pens in the compound throughout the year, and the manure they produce is transported to the fields during the dry season. Droppings of donkeys, sheep and goats are often added to the heap of household waste. However, droppings from small ruminants are sometimes used separately, for example on infields to manure certain spots in the millet crop.

Manure from small ruminants takes longer to have an effect on crop yields than cow dung, but once it has started the effect lasts for several years. As this dung is kept the nitrogen losses can be reduced by mixing with low quality biomass, i.e. with straws, leftover feeds, dry leaves, etc. The greener leaves still contain much nitrogen themselves and they are less capable of capturing surplus nitrogen from urine and dung.

Management of Household Waste and Compost

Household waste consists of partially decomposed waste from a variety of origins. It plays no part in fertility management where land is abundant. However, farmers are prepared to work hard to retain nutrients on their soil when land for cultivation becomes scarce and when food demands increase (LEIA and NCA). For example, farmers in Dilaba, Mali, said they started applying household waste in the 1970s after the drought, combining it with cow dung and the droppings of small ruminants. The system is now widely used in Dilaba and it is also starting to gain ground in the surrounding areas. The amount applied by different farmers depends largely on how much they are able to transport and on how much they can collect.

Composting is a process to break down the organic materials to make the nutrients in the biomass accessible to plants. It also helps to make a fine, easy-to-handle material and, if well done, the composting kills weed seeds. Compost is normally produced from a mixture of all kinds of organic waste, including crop residues and household waste. The technology is very ancient and there are many different ways to compost organic materials. The most common method uses a moist heap with enough nitrogen for the breakdown of the organic material that is regularly turned to allow air and oxygen to enter the heap.

Prevention of Nutrient Losses

Different kinds of nutrient losses can occur during crop cultivation and post harvest:

— Stray animals can damage the standing crop during crop cultivation and post-harvest period, in terms of losses of nitrogen, phosphorus and potassium and in terms of feeding value.

— Straw stubble that is left on the field can be trampled or soiled and incompletely eaten: careful harvest could save valuable nutrients.

— Leaving the crop residues in the field causes a loss of nutritive quality and sometimes leaching of nutrients through rains and degradation processes that involve fungi.

— Lush regrowth, in some cases after sudden rain, can be poisonous, such as the formation of prussic acid in sorghum ratoon.

— When the soil is bare for long periods nutrients can be lost due to erosion.

Animals leave their droppings while they graze the stubble, so even farmers without livestock will receive some on their fields. However, the quality of the dung is likely to deteriorate if left lying on the field unprotected for up to six months, even when it is stacked and stored in the open. Grazing animals do not "produce" nutrients in dung and urine; actually they remove more biomass and nutrients from cropland than they return in the form of manure, as shown by studies in some West African countries (Table 3).

Activities to prevent or reduce nutrient losses are:

— Penning of animals with or without bedding.

— Storage of crop residues.

— Mulching of the soil with residues unfit for animal feed.

— Breeding or managing for crop varieties with better or more straws or for cultivars that have more ratoon.

Table 3. Cereal stover removals and manure returns during crop residue grazing in West Africa

Location	*Total (kg/ha)*	*Dry*	*Matter (kg/ha)*	*Nitrogen (kg/ha)*	*Phosphorus*
		Stover removal			
Nigeria	2 470			24.5	3.9
Burkina Faso	1 570			14.8	4.0
Niger	2 135			16.8	1.8
		Manure returns			
Nigeria	27-272			0.3-1.7	0.1-0.3
Burkina Faso	600-1 600			7.5-20.0	1.5-4.0

— Covering of soil with so-called catch crops that retain leached nutrients to build their own organic matter that can be used for animal feed or mulch.

— Harvesting of crop residues before they overmature

Cropping Patterns, Livestock and Nutrients

The survival of farmers in mixed systems depends to a large extent on a proper adjustment of soils, plants and animals. Techniques include:

— better use of fallow;

— use of leys and catch crops.

These techniques can help in several ways either to reduce erosion, to add nitrogen, to enhance phosphorus flow rates, to pick up leached nutrients or to provide feed.

Use of Fallow

Farmers in most EXPAGR regions tend to depend on fallow as the technique for restoring soil fertility. This is possible because there is plenty of agricultural land available and entire fields can be left fallow for long periods of ten years or more. The system resembles shifting cultivation and the location where crops are cultivated changes every couple of years. However, over the years, with increasing population and changing technology, the fallow period has become shorter and, in areas with a dense population, it has disappeared almost entirely.

Use of Leys and Catch Crops

A ley is a fallow that is planted with crops such as grass or legumes to regenerate the soil more rapidly. The grass is used as fodder if animals are present on the farm, but similar systems are used when only mixing between crops takes place. In Mali, for example, farmers grow millet and cowpea on the same infield year after year. When they find that the fertility of part of the plot has fallen significantly, they change the association to pure stands of groundnuts or Bambara groundnuts. The following year they apply organic fertiliser and then start cultivating millet/cowpea again. Such a strategy is widely used in the Sahel.

In other words, a ley is a crop rotation that includes animals. To some extent it is also the system used in the Gangetic plains and in the Nile delta, where berseem and/or mustard are rotation crops used exclusively for livestock feeding, either on-farm or off-farm. The system is being tried in northern Thailand, and is also practised in the Krishna Valley of eastern India where a legume is the main rotation crop. However, use of legumes is also common in the United States, Western Europe and in Mediterranean systems that even occur in Australia.

Integration of crops and livestock is often considered as a step towards sustainable agricultural production because of the associated intensified organic matter and nutrient cycling. Residues of the different crops represent the main on-farm source of organic matter and nutrients. This combines well with the presence of livestock since animals play a vital role as capital assets for security and as a means of saving, for cash income and in nutrient flows. Management of crop residues in such regions is closely related to their utilisation in animal feeding.

An advantage of the integration of livestock and crops is the added value derived from crop residues in terms of animal products and income. However, to maintain the system in the long term it is necessary that nutrients from external sources are added.

In the past, this was covered to some extent through fallowing and manuring contracts with pastoralists, but the growing demand for feed by the increasing herd, the shrinking area of cropland per capita and declining crop yields dictate the scope for improvement through integration of crops and livestock.

The extra input must come from inorganic fertiliser as well as from concentrates or both. Often, the price ratios of fertiliser and grains are not conducive to the utilisation of fertilisers, and development of institutional and physical infrastructure for cost-effective use of fertilisers and concentrate is required to trigger sustainable land use in the region. Use of concentrate can be remunerative for farmers but credit facilities may stimulate intensification of livestock production and thereby increase availability of nutrients for crop land.

Energy, Biogas and Nutrients

Animals can also play a role in the provision of energy. Sometimes this role is very negative where livestock keeping contributes to deforestation in large parts of South America, Asia and Africa, but it can also be positive, such as by transforming plant energy into useful work or by providing dung used for fuel through dungcakes or biogas to replace charcoal. Wood and charcoal will con-tinue for a long time to provide energy for household cooking and for rural industries but increasing energy demands and increasing pressure on natural resources call for measures such as a more efficient use of biomass, improved natural resource management, fuel switching from, for example, charcoal to kerosene or liquefied petroleum gas, and for the development of alternatives such as biogas. The principal uses of biogas are for household energy, such as cooking and lighting, although larger installations can produce sufficient gas to fuel engines, for example, for powering mills and water pumps.

Despite the merits of biogas technology, bio digesters have only been widespread in India and in China . In these countries it is common to find biodigesters even in remote villages provided there is enough water. For example, an ambitious programme in Nepal is accelerating market development of small biodigesters. In many countries, some practical experience has now been gained with the dissemination of biodigesters but larger-scale dissemination has not yet become popular, despite the assumed economic, environmental and social benefits of the technology.

The main reasons seem to be that the multiple benefits of biodigesters must be acknowledged by the end-users to convince them to invest in the technology, and that a market-driven institutional infrastructure should be in place to facilitate large-scale dissemination of the technology. Much work has been done on mixed crop-livestock systems including the use of biogas, for example in the well-known CIPAV project in Colombia where pigs, sheep, sugar cane and biogas are just some of the components studied.

Three designs can be distinguished among the small-scale and low-cost biodigesters:

— the floating-drum, also known as the "Indian design";
— the fixed-dome, known as the "Chinese design"; and
— the flexible-bag digester.

Roughly speaking, a well-constructed fixed-dome biodigester has the longest lifetime: 20 years and more. The floating-drum digester can have a comparable lifetime, but the recurrent costs are higher as the steel drums have to be replaced every five to ten years due to corrosion. The lifetime of the flexible-bag digester is hard to indicate, as some do not last more than a couple of months, while others function for several years. Complex microbiological processes take place within a biodigester. In a common farm-type digester, the temperature ideally varies between 30 and 40 degrees Celsius. The process can be divided into three steps.

During the hydrolysis phase, the complex molecules in the feedstock are broken down into smaller molecules. During the second phase, volatile fatty acids are formed. Finally, during the methanogenic phase, other bacteria use the acids to produce methane. The methanogenic bacteria can only do their work under anaerobic conditions, i.e. it is important for a biogas plant to be built of leak-tight materials. In practice, the three phases of the process take place simultaneously in a digester where the different bacteria work together in a symbiotic relationship.

Many different types of organic matter can be used to "feed" a biodigester, but to allow the bacteria to do their work the organic matter should be accessible to them. This means that pretreatment is sometimes necessary, e.g. through chopping and/or composting of crop wastes. The easiest feedstock to use is cattle dung, as it already contains the right bacteria and the vegetable matter has been broken down by passing through the guts of the cow. Human excrement and manure from chickens and pigs are also useful, but they do not contain the right bacteria and they need a starter in the form of, for example, some slurry from a working biodigester. Neither the feedstock nor the water should contain toxins such as antibiotics, detergents and disinfectants.

Toxins cause the digester to "go sour", causing it to give off an unpleasant smell of acids and, therefore, the dung of cows that receive antibiotics should not be fed into the digester. Cultural aspects might influence the selection of feedstock. For example, Hindus in India usually accept cow dung for use in biodigesters, while pig dung and human excrement are often not accepted, particularly in Moslem regions.

Initially, biodigesters were promoted only for their energy production. Once again, the thinking about a technology in a mixed system focused on only one aspect. Unfortunately little attention has been paid to the excellent properties of the slurry as an organic fertiliser of crops and trees. In other words, slurry has the potential of supporting crop production and, in addition, the process kills or substantially reduces the amount of pathogenic germs and seeds of weeds present in the feedstock.

Another advantage is that nutrients that are already present in the feedstock, such as nitrogen, potassium and phosphorus, are made available to crops more quickly and efficiently, i.e. the biodigester affects the nutrient dynamics of the system. Further, it reduces the odours given off by manure and compost. Emissions of greenhouse gases are also reduced, as the biodigester produces renewable energy and because fossil fuel based chemical fertiliser is replaced by an organic fertiliser.

A major disadvantage of the slurry is that it is liquid, with a dry matter content of less than 10 percent. Transport to the fields is therefore often complicated and the use of the slurry is usually limited to fields near the digester. Larger installations in industrialised countries separate the slurry into a liquid and a solid fraction; the solid fraction can be applied on fields at a distance from the digester.

Agroforestry and Soil Fertility

Alley cropping is an agroforestry system particularly for the humid or subhumid areas, developed at the International Institute of Tropical Agriculture (IITA), Nigeria, but based on earlier farmer experience (ITK). In this system food crops are grown in alleys formed by hedgerows of trees and shrubs, preferably legumes. In most tests and projects the fast-growing species *Leucaena leucocephala* and *Gliricidia sepium* have been used. The hedgerows are pruned regularly, thus providing fodder and fuelwood, and/or mulch. They are pruned to reduce shading for the crops sown in the alleys and the approach is considered to be an improved bush-fallow system permitting continuous cropping, among other reasons, because the tree is a giant "catch crop" that recycles nutrients leached to deeper soil levels.

The alley cropping system can sustain or even improve soil fertility and crop production. The system, like any other mixed system, produces tree products such as firewood, stakes for construction wood and fodder for livestock. The latter is mostly used by cattle and small ruminants. Alley cropping is being developed with different tree species and crops, focusing on several uses of the by-products from the trees.

Alley cropping as such is not the solution to all soil fertility problems. Like other technologies it works only under certain climatic and economic conditions. Disadvantages are that it requires much labour and that feed cannot always be delivered on time. It also takes up space, i.e. it is again a reflection of the communal ideotype serving several goals and adjusting to other components in the system. A more intensive form of alley cropping is that of agroforestry in humid lowland tropics in a broad range of systems, some of which are traditional, such as multistorey gardens.

Semi-arid areas have low rainfall and tree densities are therefore lower, i.e. windbreaks can serve as a variation on the theme of agroforestry, for example in the Niger with neem trees. In highland agriculture, together with grasses and shrubs, trees can be part of an erosion control system, as found in the Rwandan contour ploughing example that uses *Grevillea robusta* trees on the contours.

3

Analysis of Non-GM Biotechnologies

In developing countries, a number of non-GM biotechnologies are currently being used in the agricultural sector. These techniques have been grouped in four main clusters: microbial, cell biology, molecular marker and diagnostic, each of them divided into subgroupings for specific techniques. Subgroupings under "other" have been included to cover unspecified activities under each cluster.

Microbial Techniques

Microbiological techniques are central to the production, harvesting and use of many agricultural products. The use of microbial techniques includes the design or delivery of microorganisms for the control of pests, as fertilisers and for fermentation and food processing techniques. From data in the FAO-BioDeC (Table 1), it seems that research towards the scientific development of biocontrol agents (biopesticides), biofertilisers and fermentation and food processing microbial techniques, is at the early stages in Africa and Asia, while Latin American countries are already using advanced techniques on a more routine basis and some of the results obtained are already being tested. It should be borne in mind that many production and post-production processes already in use by farmers have a microbiological basis and that the extent of use of microbiological techniques in agriculture is grossly under reported by focusing only on the use of microbiological techniques which emanated from 'modern' laboratories, rather than also on the use of microbiological techniques developed as indigenous or local knowledge of farmers and consumers.

Microbial Agents in Pests Control

Although considerable use of classical pesticides persists, in some countries there is a trend towards using newer pesticides that are more selective, less toxic to humans and the environment, and require less application per hectare to be effective. A small but growing percentage of these are biopesticides, including microbial pesticides like *Bacillus thuringiensis* (Bt), and biocontrol agents such as pheromones, growth regulators and hormones. In addition, there is growing acceptance of use of such alternative pest control agents in various forms of Integrated Pest Management (IPM). Biocontrol agents, or biopesticides, range from the classic *Bacillus thuringiensis* (Bt) to *Trichoderma, Verticillium, Bauveria, Bacillus subtilis* to plant extracts, entomophagic nematodes or entomopathogenic viruses, such as nucleopolyhedrosis virus (NPV).

Table 1 shows that reported research on biological control agents is only at the laboratory phase in Africa, with emphasis on application to cowpea in Ghana and Kenya, sorghum in Ethiopia, Kenya and Zimbabwe, banana in South Africa, Uganda and Zimbabwe, cassava in Malawi and sugar cane in South Africa. In Eastern Europe, both Georgia and the Republic of Moldova have

Table 1

Number of initiatives to develop and use microbial techniques

Technique	*Region*																													
	Africa					*Asia*					*Eastern Europe*					*Latin America*					*Near East*					*Total*				
	C	*T*	*E*	*U*	*N*	*C*	*T*	*E*	*U*	*N*	*C*	*T*	*E*	*U*	*N*	*C*	*T*	*E*	*U*	*N*	*C*	*T*	*E*	*U*	*N*	*C*	*T*	*E*	*U*	*N*
Biocontrol	-	-	11	-	7	-	-	3	-	3	-	2	6	-	2	-	9	2	-	2	-	-	2	-	2	-	11	24	-	16
Bio-fertilizers	1	-	7	6	10	-	-	10	3	6	-	4	1	-	4	-	3	3	-	3	-	-	1	-	1	1	7	22	9	24
Ferment-ation	-	-	4	-	5	-	-	7	-	2	10	1	1	-	4	1	3	7	-	5	-	-	-	-	-	11	4	19	-	16
Other	-	-	-	-	-	-	-	-	3	2	-	3	3	-	2	-	2	5	-	4	-	-	-	-	-	-	5	8	3	9
Total	1	-	22	6	12	-	-	20	6	7	10	10	11	-	5	1	17	17	-	6	-	-	3	-	2	12	27	73	12	32

trials underway on use of *Trichoderma* spp as a biocontrol agent. Research on *Metarhizium anisopliae, Entomophora* spp. is undertaken in Georgia and on pathogenic bacteria, fungi and viruses in the Republic of Moldova.

In Asia, use of *Steinernema thailandensis* is established and laboratory research is underway on Bt and NPV for use as biopesticides in Thailand. In Latin America, Guatemala is testing botanical extracts and *Bacillus subtilis* on melons and tomato, *B. subtilis* on pea, entomopathogenic fungi on sugar cane and pasture, and Bt/NPV on ornamental, horticultural and other crops. Chile is researching biocontrol of scab in apple. In the Near East, Egypt and Morocco are researching biocontrol of Fusarium in date palm.

Biofertilisers

Biological nitrogen fixation (BNF) refers to the process of microorganisms fixing atmospheric nitrogen, mostly within subsoil plant nodules, and making it available for assimilation by plants. Nitrogen availability is a key limiting factor in crop production. *Rhizobia* are the most studied and important genera of nitrogen fixing bacteria, but also a number of endophytic bacteria are now known as nitrogen fixing organisms. Use of biofertilisers, such as *Rhizobium* or other symbiotic and non-symbiotic species for atmospheric nitrogen fixation represents a more environmentally-friendly alternative to chemically generated fertilisers. Other microorganisms, such as *Mycorrhiza*, are active in establishing symbiosis with cultivated plants and forest trees, and facilitate phosphorus uptake. Inoculation with these fungi has proven to be an efficient way to substitute or complement phosphorus-based chemical fertilisation.

The only reported research on biofertilisers in the Near East is research into *Rhizobium* in food legumes in

Morocco. In Latin America, techniques for increasing nitrogen fixation in soybean, bean and cowpea are being tested in Venezuela, and researched in rice in Brazil, where arbuscular mycorrhiza are also being tested in native trees. Argentina has started nitrogen-fixation research on unspecified microorganisms. In Asia, India is researching *Rhizobium* with high capacity for nitrogen fixation, and China is studying *Rhizobium* in rice and maize. Indonesia is investigating vascular arbuscular mycorrhiza, the Philippines mycorrhiza and *Rhizobium* in forest species, and Viet Nam is experimenting with *Rhizobium* strains for the Mekong delta soils.

In Africa, research into biofertilisers for sorghum is ongoing in Ethiopia, Kenya and Zimbabwe, for cowpea in Cameroon, for groundnut and bambara groundnut in Madagascar, and for rice in Rwanda, with unspecified work in Burkina Faso, Cote d'Ivoire, the Democratic Republic of the Congo, Kenya, Rwanda and Senegal. The UNESCO Microbiological Resources Centre (MIRCEN) project at the University of Nairobi in Kenya has, since 1981, developed a *Rhizobium* inoculant known as *BIOFIX*, currently the main inoculant available on the local market. In Eastern Europe, Armenia and the Republic of Moldova are testing *Azotobacter* and *Rhizobium*.

Fermentation Technology and Food Processing Techniques

Biotechnology in the food processing sector targets the selection and improvement of microorganisms with the objectives of improving process control, yields and efficiency as well as the quality, safety and consistency of bioprocessed products. These microorganisms are used in fermentation; the process of bioconversion of organic substances by microorganisms and/or enzymes of microbial, plant or animal origin. Fermentation is one of the oldest forms of food processing which is applied globally. In developing countries, fermented foods are produced primarily at the household and village level, where they find wide consumer acceptance. Food fermentations contribute substantially to food safety and food security, particularly in the rural areas of many developing countries.

The extent of application of biotechnologies such as fermentation technology and various food processing techniques is immense and difficult to monitor using a database such as the FAO-BioDeC. Nevertheless, it indicates that fermentation technology and food processing is quite commercialised in Latin America for sorghum in Brazil, where food enzymes are also being tested in several species of grains and vegetables, and researched in sugar cane. In Eastern Europe, *Lactobacillus*, *Acetobacter* and *Saccharomyces* are reported as being used commercially in Armenia, Georgia and the Republic of Moldova, where *Bacillus subtilis*, is also being used. *Actinomycetes* are being tested in Azerbaijan, and *Bacillus vulgaris* researched in Armenia.

In Africa, research phase initiatives are underway in the fermentation of banana in Kenya and Zimbabwe, cowpea in Cameroon, cassava in Nigeria, with unspecified research in Burkina Faso. In Asia, Indonesia is researching lactic acid bacteria, dextranase, xanthan gum production and soybean fermentation. China and Indonesia are reported to be working on the use of the phytase enzyme in improving animal feed.

Development of Environmental biotechnology

Environmental biotechnology can be defined as the development, use and regulation of biological systems for remediation of contaminated environments (land, air, water), and for environment-friendly processes. The FAO-

BioDeC reveals few environmental biotechnology initiatives of relevance to food and agriculture underway in the countries surveyed. However, it is reported that microbiologial techniques for environmental management are being investigated in Indonesia, for biological waste and water treatment and bioconversion of solasodine, and in Thailand for biodegradable plastics and waste treatments. In Eastern Europe, Azerbaijan is testing *Basidiomycetes, Deyteromycetes,* and *Fungi imperfecti* and Georgia is researching *Mycobacteria, Haiobacterium halovium,* and *Aspergillus terreus.* In Latin America, Chile is testing *Thermomyces lanuginosus* for cellulase production, and bacterial biofilm for bioremediation.

Applied Plant Cell Biology Techniques

Micropropagation

Micropropagation is the use of tissue culture methods to propagate plants. Using micropropagation, millions of new (clonal) plants can potentially be derived from a single plant. Micropropagation encompasses a range of tissue culture techniques for propagation of plant species. In essence, tissue from a plant (explant) is isolated to create a sterile culture of that species *in vitro.* Once a culture is stabilised and growing well *in vitro,* multiplication of the tissue or regeneration of entire plants can be carried out. Shoots (tips, nodes or internodes) and leaf pieces are commonly used but cultures can be generated from many different tissues. Juvenile tissues generally respond best. This method of cultivation of plant material is used for:

— rapid, large-scale, year round production of desired horticultural varieties (e.g. many of the popular miniature rose varieties are produced this way);

— propagation of plant species that are difficult to grow from seed;
— production of genetically uniform plant material ("clones");
— development of plant culture systems that can be used for genetic transformation, e.g. to introduce disease resistance; and
— production of disease-free plant material (e.g. potato microtubers).

The benefits of plant tissue culture propagation include potentially unlimited multiplication of superior plant lines or elite individuals, avoidance of contamination with pathogens, production of true-to-type multiplication material of desirable plant lines suited for indefinite storage through cryopreservation, or for long-term maintenance of propagule inventories. In comparison to other plant propagation techniques, the major limitation of the application of this technology is the need for technically skilled labourers and some essential equipment. It is therefore commercially applied to high value added crops, which are worth the necessary investment.

Micropropagation is now a 'mature' plant biotechnology and is among the most widely used plant biotechnologies, reportedly being applied in 21 countries in Africa, ten in Asia, nine in Eastern Europe, nine in Latin America and eight in the Near East. The success of micropropagation may be explained by its relatively low costs and generally positive effects on productivity. Micropropagation has become an irreplaceable tool for many clonally propagated species for the production of pathogen-free plantlets. Emphasis on the development of cost-effective micropropagation techniques is expected to increase even further.

The FAO-BioDeC indicated that micropropagation is routinely being used, with many commercial products available on the market. In Latin America, micropropagation is used for blueberry, raspberry, sugar cane and potato in Argentina, and sugar cane, blackberry, strawberry, apple, potato, banana, cassava, pineapple, violet, fruits, cereals and ornamentals in Brazil. In Asia, Bangladesh has micropropagated plants of banana, orchid, chrysanthemum, potato, jackfruit, pineapple and unspecified timber trees; Nepal has micropropagated potato, orchids and *Miscanthus* spp. In Eastern Europe, Yugoslavia has micropropagated potato, while in the Near East, date palm is being commercially propagated in the Islamic Republic of Iran, Kuwait, Morocco, Tunisia and the United Arab Emirates. Tunisia also has available micropropagated *Prunus* rootstocks, almond, citrus, grape, olive and pistachio.

The Islamic Republic of Iran, Jordan and the Syrian Arab Republic have potatoes, and in addition banana is available in the Islamic Republic of Iran. In Africa, micropropagated banana is promoted in Cameroon, Gabon, Kenya and Uganda. In addition, plantain and cassava are promoted in Gabon, and yam, potato and sweet potato in Uganda. A wide range of micropropagated ornamentals, trees and food crops, with emphasis on root and fruit crops, are being tested for commercial production, or are in the preliminary research phase in all regions.

Anther and Pollen Culture

Anther culture involves the aseptic culture of immature anthers to generate fertile haploid plants from microspores. The production of haploid plants through anther culture is widely used for breeding purposes, as an alternative to the numerous cycles of inbreeding or backcrossing usually needed to obtain pure lines in conventional breeding. The success achieved with anther culture has led to the development of microspore culture. This involves the isolation of the microspores from the anthers, culturing them in specialised media and subsequent regeneration of fertile homozygous plants. Furthermore, isolated microspores are very attractive for protoplast isolation and applications aiming at transformation as they are unicellular and transgenic homozygous plants could be provided in a comparatively short time.

In vitro anther culture is now routinely used for improving some vegetable crops such as asparagus, sweet pepper, eggplant, watermelon and Brassica vegetables. In addition, anther culture is being increasingly used in cereal crop improvement both as a source of haploids and new genetic variation. Isolated microspore culture has been successfully carried out with some Brassica vegetables such as cabbage, broccoli and Chinese cabbage-petsai and pakchoi. Most genotypes respond better to isolated microspore culture and embryo yield is generally higher than with anther culture. Therefore, isolated microspore culture has been preferred as a breeding tool and as an experimental system for various genetic manipulations. A Chinese cabbage breeding programme using this technique is underway in the Beijing Vegetable Research Center.

In Asia, three countries are reportedly using anther culture. In China, for example anther culture is being applied in different species such as wheat, rice, maize, rubber, hot pepper, poplar and sugar beet. Bangladesh is conducting trials on anther culture derived rice with salt tolerance and anther culture research on mulberry, maize, jute barley and tea, and Nepal has research on rice and wheat. In the Near East, anther culture research emphasis is on cereals.

Table 2. Some examples of countries using plant propagation techniques for different crops at the commercial level

Crop or species	*Countries*
Micropropagation	
Potato	Bangladesh, Islamic Republic of Iran, Jordan, Nepal, Syrian Arab Republic, Uganda.
Sweet potato	Uganda
Plantain, cassava	Gabon
Banana	Bangladesh, Cameroon, Gabon, Islamic Republic of Iran, Kenya, Myanmar, Uganda
Date palm	Islamic Republic of Iran, Kuwait, Morocco, Tunisia, United Arab Emirates
Jackfruit	Bangladesh
Pineapple	Bangladesh, Myanmar
Citrus, almond, *prunus* rootstocks, grape, olive, pistachio	Tunisia
Anther culture	
Bread wheat	Morocco
Durum wheat	Tunisia
Fruits, cereals and ornamentals	Brazil
Embryo rescue	
Banana, citrus, papaya, fruits, cereals and ornamentals	Brazil
Germplasm conservation	
Cassava, violet	Brazil
Other propagation techniques	
Kakrol	Bangladesh

An anther culture generated durum wheat variety has reached the commercial phase in Tunisia, and anther culture techniques are being researched in Morocco, where an anther cultured bread wheat is in commercial use. Anther culture techniques are being tested in the Sudan, and applied to barley in the Islamic Republic of Iran, Iraq, Morocco and the Syrian Arab Republic. Wheat is also being researched regarding anther culture in Eastern Europe in the former Yugoslav Republic of Macedonia and Serbia and Montenegro.

In Latin America, varieties obtained through anther culture of unspecified fruits, cereals and ornamentals have reached the commercial stage in Brazil. Advanced testing of anther culture derived varieties of 'many plants' is reported in Costa Rica, and in Venezuela cocoa, rice, sugar cane, papaya, mango and bamboo are also being tested. Anther culture derived rice varieties are under advanced testing also in Argentina, and in Brazil, where this technique is specifically used in breeding for blast resistance. In Chile, wheat, barley and rice, are in the research phase. Other anther culture research is carried out on maize and onion in Argentina, wheat in Ecuador and asparagus in Peru. In Africa, anther culture research has started on banana in Cameroon and on sorghum and rice in Mali.

Embryo Culture Techniques

In embryo rescue/culture, the embryo is removed before seed abortion occurs and is grown outside the parent plant to produce a new plant to enable crosses to be made between species which would not normally be sexually compatible. *In vitro* embryo rescue techniques are therefore often used to rescue plant embryos from aborting progeny seeds that result when two distantly related plants (e.g. two species) are crossed together. Such 'wide crosses' are often desirable to transfer genetic traits from the secondary and tertiary genepools (i.e. crop wild relatives) to the cultivated primary genepools of crop plants. This technique is used in breeding of many crop species and allowed synthesis of triticale, a new hybrid species resulting from the cross between rye and wheat.

In the Near East, embryo rescue techniques are being developed in Algeria, Morocco and the Sudan. In Eastern Europe, embryo culture research is underway on sweet and sour cherry in the former Yugoslav Republic of Macedonia, sunflower in Yugoslavia, and maize in the Republic of Moldova. Only in Latin America has extensive use been made of the embryo culture technique, where it has given rise to commercially viable new varieties in banana, citrus, papaya, fruits, cereals and ornamentals in Brazil. New embryo rescue-derived varieties are tested in maize, *Celtis tala, Erthytrina crista galli, Acacia caven* and *Elymus spp.* in Argentina, in grape in Chile, in orchids in Costa Rica and in rice, sugar cane, papaya, *Solanum nigrum,* Yantén, *Spathiphyllum* sp., *Aster* sp., and gloxinia in Venezuela. Research is at an early stage in Africa, on yam in Nigeria and unspecified work is being carried out in Cameroon, and also in Asia, where the only embryo culture report is of work on rice and orchid in Bangladesh.

In Vitro Regeneration and Somaclonal Variation

Plant regeneration from cell cultures is central to the application of gene transfer techniques such as biolistics and *Agrobacterium*-mediated transformation. Not all plants are readily amenable to *in vitro* regeneration and there is thus a need to continuously develop regeneration protocols for the recalcitrant species if they are to benefit from genetic modification technologies. In vitro regeneration usually results in high genetic and phenotypic variability in individuals derived from cultures, which is called somatic variation. Somatic variation can be beneficial in crop improvement especially on traits for which somaclonal mutants can be enriched during *in vitro* culture, including resistance to disease pathotoxins, herbicides and tolerance to environmental or chemical stress, as well as for increased production of secondary metabolites.

Advanced *in vitro* regeneration work (protocol testing) is reported in *Acacia caven, Erythrina crista galli*, potato, sweet potato, cassava and garlic in Argentina, in blueberry in Chile, in cassava, aroids, *Musa* sp., *Coffea* sp., banana and plantains in Costa Rica, in *Ullucus tuberosus, Oxalis tuberosa, Tropaeolum tuberosum, Arracacia xanthorrhiza, Mirabilis expansa, Canna edulis*, and manihot in Ecuador and in tropical fruits, roots and tubers, cassava, potato, *Musa, Solanum nigrum*, Yantén, *Spathiphyllum* sp., *Aster* sp., and gloxinia in Venezuela. Research is in the laboratory stage (protocol development) on *Parkinsonia aculeate, Elymus* sp, and *Ilex paraguariensis* in Argentina, red clover and garlic in Chile, and on *Mycosphaerella fijiensis, Rosellinia* spp. and selected plant species in Costa Rica. In Asia, the only reported initiative is research on somaclonal variation in mulberry in Bangladesh.

In Vitro Germplasm Conservation and Cryopreservation

Biotechnology can play an important role in the conservation of germplasm of many crop species, whether the germplasm is maintained as vegetative or seed propagules. *In vitro* conservation protects germplasm from possible contamination with pathogenic agents and preserves the genetic identity of the stored material. Germplasm regeneration techniques (*in vitro* or *in vivo*) coupled to cryopreservation protocols ensure the long-term, safe storage of much of the world's germplasm, even if this technique is well established only for a number of plant species. The FAO-BioDeC include information on the use of *in vitro* conservation and cryopreservation. In this regard, the only work reported in Asia is experiments in Viet Nam on sugar cane and in Bangladesh on orchid, bamboo and hybrid *Acacia*.

Similarly in Africa, reported work is only in the experimental phase, concentrating on root crops, cassava in Cameroon, Ghana and Nigeria, yams in Cameroon, Malawi and Nigeria, sweet potato and potato in Kenya, also on banana in Ghana, Malawi, Nigeria and Uganda, with unspecified activities in Cote d'Ivoire. In Latin America, use in cassava and violet is reported in Brazil, where experiments are also underway on sugar cane. Chile has developed cryopreservation of *Fragaria chiloensis* at the laboratory stage. In the Near East, only Egypt has any activity, in the laboratory, on potato conservation.

In Eastern Europe, the former Yugoslav Republic of Macedonia is conducting trials on techniques for *in vitro* conservation of grape, apple, strawberry and cherry, and Yugoslavia for sunflower and potato. Experimental work is ongoing for potato in Albania and Bosnia and Herzegovina, also on *Populus, Genthiana* sp., and 30 wild species in Bosnia and Herzegovina and on *Salvia hydrangea, Bryonia alba, Atropa belladonna, Rubia tinctorum, Hypericum perforatum,* and *Melissa officinalis* in Armenia and grapevine in Azerbaijan.

Molecular Marker Techniques

Molecular marker techniques represent a rapidly evolving suite of powerful research tools for the characterisation and management of genetic polymorphism in plant breeding and germplasm characterisation programmes. Where they are available and cost-effective, molecular markers can have a wide number of applications in plant breeding, the most commonly considered being marker-assisted selection (MAS). The FAO-BioDeC suggests that the earliest generation forms of DNA-based molecular markers (RAPD, RFLP) are more widely used than the more recently developed types of molecular markers.

MAS is based on the identification and use of markers which are linked to the gene(s) controlling the trait of interest. By virtue of that linkage, selection may be applied to the marker itself. The advantage consists in the opportunity of speeding up the application of the selection procedure. For instance, a character which is expressed only at the mature-plant stage, may be selected at the plantlet stage if selection is applied to a molecular marker. Also, selection may be applied simultaneously to more than one character, and selection for a resistance gene can be carried out without needing to expose the plant to the pest, pathogen or deleterious agent.

Finally, when there is linkage between a molecular marker and a quantitative trait locus (QTL), selection may become more efficient and rapid. The construction of detailed genetic and molecular maps of the genome of the species of interest is a prerequisite for most forms of MAS. However, the current cost of the application of these techniques is significant, and the choice of one technique rather than others may be dictated by cost factors. There are still very few examples of crop varieties in farmers' fields which have been developed based on MAS, largely because of the currently prohibitive cost of incorporating large-scale MAS into the budgets of most plant breeding programmes.

In addition to MAS, molecular markers can be used in germplasm characterisation. Compared to morphological and protein markers, DNA-based genetic markers are often considered to be the most useful for genetic diversity studies because they are highly polymorphic and heritable. The FAO-BioDeC indicates that molecular markers are being extensively used in Latin America with 93 trials and 165 molecular marker projects at the research phase in nine countries.

The assessment of the genetic diversity of Andean local roots and tubers in Latin America using molecular markers is an interesting development. Species reported to be included in molecular marker programmes in these countries are sugar cane, rice, cocoa, banana, bean and maize. The survey also indicates that most countries in Asia are undertaking a wide spectrum of crop research using molecular markers. Molecular marker-related research activities in Africa are reported to be underway in only a few of the countries, such as Ethiopia, Nigeria, South Africa and Zimbabwe; the range of Africa crops under study with molecular markers, however, is very wide: from traditional commodities to tropical fruits.

DNA and Immuno-Diagnostic Techniques

Many crop diseases are difficult to diagnose, especially at the earliest stages of infection by the pathogen. In particular, many diseases caused by viruses can exhibit similar symptoms and therefore it is difficult to identify the causative virus. Knowledge of the nature of the pathogen can be used to develop and apply proper management practices. For instance, some viruses are seed transmitted

The development of cheap diagnostic techniques could assist decision-making in relation to pest and disease management. The development of diagnostic kits, such as enzyme linked immunosorbent assay (ELISA), and other molecular assays, can enable the precise identification of viruses, bacteria and other disease-causing agents, and is now an established tool in disease management in many farming systems. Diagnostic assays have also been developed to identify a wide range of other organisms, chemicals, or impurities involved with food quality.

The FAO-BioDeC (Table 3) suggests that ELISA is only in commercial use in Eastern Europe, although this is unlikely to be unique to Eastern Europe as ELISA is undoubtedly used in plantation farming in many other regions. Most of the use of ELISA reported in Eastern Europe is to detect plant viruses in crops such as potato, fruits and strawberries. In the Near East, all ELISA-related work is reported to be in the experimental phase, with Iraq, Morocco and the Syrian Arab Republic working on potato, Morocco also on sugar cane, Tunisia on grapevine and Egypt on detection of ZYMV in cucurbits.

In Africa, use of ELISA for cassava diagnostics is under study in Malawi, Nigeria, Uganda and Zimbabwe, for cowpea in Cameroon and Nigeria, and maize in South Africa and Zimbabwe. In Latin America, Brazil is testing ELISA in sugar cane, potato, plum, unspecified Solanaceae and soybean, and researching cucurbits, Solanaceae, and Rhizobium in association with beans and forage and tree legumes, Chile is researching potato, tomato, tobacco and grapevine, Paraguay is testing ELISA in citrus, and Peru is testing potato.

DNA diagnostics are also a powerful technique for identification of pathogens and other organisms in agriculture. Most DNA diagnostics are now based on the use of the polymerase chain reaction (PCR), a common research tool used in most molecular biology laboratories worldwide, which can be used to specifically amplify

Table 3

Number of initiatives to develop and use Number of initiatives to develop and use diagnostic techniques

Technique	*Region* *Africa*					*Asia*					*Eastern Europe*					*Latin America*					*Near East*					*Total*				
	C	T	E	U	N	C	T	E	U	N	C	T	E	U	N	C	T	E	U	N	C	T	E	U	N	C	T	E	U	N
ELISA	-	-	11	-	6	-	-	-	-	-	1	3	9	-	4	-	13	10	1	6	-	-	7	-	5	1	16	37	1	21
Monoclonal Antibodies	-	-	2	-	2	-	-	3	-	2	-	-	2	-	2	-	-	-	-	-	-	-	-	-	-	-	-	7	-	6
Nucleic acid	-	-	-	-	-	-	-	-	-	-	-	-	-	-	-	-	-	-	-	-	-	-	1	-	1	-	-	1	-	1
PCR	-	-	5	2	7	-	-	2	-	1	-	-	10	-	4	-	4	12	-	3	-	-	2	-	1	-	4	31	-	16
Other	-	-	10	-	8	-	-	2	4	5	-	-	-	-	-	-	-	1	-	1	-	-	2	-	1	-	-	15	4	15
Total	-	-	28	2	10	-	-	7	4	7	1	3	21	-	6	-	17	23	1	7	-	-	12	-	6	1	20	91	5	36

segments of DNA. Most PCR techniques require the use of the enzyme *Taq* polymerase which until recently was protected by patents requiring that any commercial use of the PCR technique would have to pay royalties to the holders of the *Taq* polymerase patent. The enzyme *Taq* polymerase is now in the process of becoming a 'generic' biochemical reagent which will substantially reduce the cost of PCR applications in research and commerce.

At present, most of the reported work on PCR diagnostics is in the experimental phase. In the Near East, Egypt is working on PCR analysis of ZYMV in cucurbits and TYLCV in tomato. In Europe, Yugoslavia is researching PCR for PPV in plum, ASPV in apple and PVYV in pear. Armenia is working on diagnostic techniques for disease of tobacco, potato, linen, wheat and other crops, Azerbaijan on chickpea, tobacco and other crops, and the Republic of Moldova on maize, tobacco and other crops.

In Africa, maize disease diagnostics is the subject of PCR studies in South Africa, Uganda and Zimbabwe, while diagnosis tools are studied for diseases of cowpea in Cameroon and Nigeria, cassava in South Africa and unspecified work is underway in Burkina Faso and Ghana. In Latin America, Brazil is researching on PCR for unspecified uses in sugar cane, bean, rice, tomato, carrot, unspecified Solanaceae and banana and is conducting experiments on sweet potato, garlic, apple, genome sequencing of *Herbaspirillum* in soybean, *Rhizobium* in association with beans and forage and tree legumes, peanut and wild related species, studies on fungi, eucalyptus, Brazilian forest trees, soybean, coffee, sugar cane, *Xylella fastidiosa*, and heart of palm for food quality. Chile is researching molecular diagnostics for pathogens of potato, tomato, tobacco and grapevine, and Peru of potato.

4

Analysis of GM Crop Varieties

The transgenic plant varieties which are resistant to pathogens, pests, herbicides, tolerant to abiotic stresses, and with modified quality traits.

Pathogen Resistance

The data contained in the FAO-BioDeC related to transgenic varieties resistant to pathogens are summarised in Table 1. These data suggest that most pathogen-resistance development programmes oriented towards the development of genetically modified (GM) plant varieties, are concentrated on generating resistance to viruses and fungi, particularly in Asia and Latin America, with very little activity devoted to developing transgenic crop varieties resistant to bacteria.

The current emphasis on viruses most likely reflects the relative simplicity of viral genomes, and the generic 'proof of concept' of virus-derived strategies such as coat protein and replicase-mediated resistance in generating virus-resistant transgenic varieties. Transgene-mediated resistance strategies against fungi are still in their infancy. Overall, it is unclear from the available data in the inventory why transgenic approaches are more

Table 1

Number of initiatives to develop GMOs with resistance to pathogens

Trait	*Region*																													
	Africa					*Asia*					*Eastern Europe*					*Latin America*					*Near East*					*Total*				
	C	T	E	U	N	C	T	E	U	N	C	T	E	U	N	C	T	E	U	N	C	T	E	U	N	C	T	E	U	N
Virus	-	2	3	-	2	2	17	14	9	9	-	-	-	-	-	-	16	26	-	9	-	9	2	-	3	2	44	45	9	23
Bacteria	-	-	-	-	-	-	1	3	5	5	-	-	-	-	-	-	-	1	-	1	-	-	-	-	-	-	1	4	5	6
Fungi	-	1	1	-	1	1	1	6	-	5	-	-	3	-	1	-	6	16	-	8	-	-	-	-	-	-	8	26	-	15
Other	-	-	-	-	-	-	-	7	-	4	-	-	1	-	1	-	3	8	-	4	-	-	-	-	-	-	3	15	-	8
Total	-	3	4	-	3	3	19	35	14	9	-	-	4	-	1	-	25	51	-	9	-	9	2	-	3	2	56	96	14	25

predominant for fungal compared to bacterial pathogens, or why some regions report more research initiatives than others.

Although planting disease-resistant varieties is one of the better ways of combating viruses, bacterial and fungal plant pathogens, as with all biotic stresses with the capacity to mutate and evolve, it would be prudent to develop resistance management strategies to limit the incidence of selection for resistance-breaking pathogen isolates.

Transgenic Crop Varieties Resistant to Viral Diseases

In the case of virus resistance, conventional strategies to control viral diseases are limited to the production of virus-free propagation material and to the control of insects transmitting virus pests. While some crop gene pools harbour resistance to viruses, there are crop gene pools which are completely lacking in resistance against key virus pathogens. The figures in Table 1 suggest a rapid adoption in some regions of pathogen-derived transgene-mediated virus resistance strategies, which were first demonstrated in greenhouses in 1986 for tobacco mosaic virus (TMV). The few examples of large-scale plantings of virus resistant GM varieties should be monitored to determine if resistance breaking variants of the virus pathogens are being selected over a number of growing seasons for large scale crop populations.

In Africa, only three crops, sweet potato, potato and maize, have so far been targeted for transgene-mediated virus resistance. The FAO-BioDeC indicates only two transgenic varieties that to date have been tested in a field trial, namely, a sweet potato variety for resistance to sweet potato feathery mottle virus (SPFMV) in Kenya and a potato variety for resistance to potato leaf roll virus

(PLRV) in South Africa. The FAO-BioDeC currently reports only three other research initiatives in an experimental phase in South Africa, namely, the development of potato for resistance to potato virus Y (PVY) and potato virus X (PVX), and development of maize resistant to the maize streak virus (MSV). The inventory indicates that no virus resistant GM varieties have been commercially released in Africa.

In the Eastern and Central Africa region, maize is a major staple of the rural and urban poor. In the same region, potato has become a major highland cash crop and a food staple in some urban areas. Sweet potato is an important crop in the countries surrounding Lake Victoria. The FAOSTAT database indicates that the 2003 production in Africa for sweet potato was 10 787 127 tonnes, potatoes 12 530 119 tonnes and maize 43 522 313 tonnes. Indigenous to Africa and its offshore islands, MSV causes yield losses of up to 100 percent even in high potential agricultural zones. MSV is considered the most serious virus threat to maize production in Africa, while SPFMV is a serious constraint to sweet potato production.

In the Near East region, Egypt is conducting field trials on virus-resistant GM varieties of potato, tomato, cucumber, melon, muskmelon, cantaloupe, squash and sugar cane. Tunisia has initiated virus-resistant GM research work on potato and *Vitis,* and The Islamic Republic of Iran has initiated this research work on sugar beet. Most virus-resistant GM varieties under field testing are of imported origin. However, one country in this region, Egypt, is carrying out field testing of locally developed virus resistant varieties, mainly cucurbits with resistance to ZYMV, and potato resistant to PVY and PLRV.

At least two virus-resistant transgenic crop varieties have been commercially released in the Asia region.

These include virus-resistant tomato and green pepper varieties in China. In general, current research efforts to develop transgenic virus-resistant crop varieties in this region are mostly focused on *Solanaceae, Cucurbitaceae* and tropical fruit trees. Special attention is being given to development of virus-resistant papaya, as transgenic papaya varieties resistant to papaya ringspot virus (PRSV) now provide an additional strategy to restore papaya cultivation in areas where the virus has been most destructive. Research has been reported in Malaysia, the Philippines and Thailand on the development of different ring spotvirus resistant transgenic varieties of papaya, some being combined with delayed fruit ripening.

Field trials are being conducted in China for transgenic varieties exhibiting resistance to: CMV in sweet pepper, CMV and TMV in chilli pepper, TMV in tobacco, PRSV in papaya, stripe virus in groundnut, PVY in potato, BYDV in wheat and turnip mosaic virus in Chinese cabbage. Laboratory stage research work on RDV in rice and WYMV in wheat is in progress in China. Thailand has field trials in progress for transgenic virus resistant varieties of tomato (TYLCV), papaya (PRSV) and pepper (CVbMV), and laboratory stage work on pepper (PepLCV), yard long bean (aphid-borne mosaic virus) and rice (ragged stunt virus).

The Philippines has field trials underway for virus-resistant transgenic varieties of banana (BTV), and laboratory stage experimental work on transgenic papaya (PRSV). Indonesia has laboratory research underway on the development of virus-resistant peanut (peanut stripe virus), tobacco (TMV), sweet potato (SPFMV), chilli pepper, papaya and potato (PVX and PVY) varieties. Malaysia is conducting transgenic experiments on the development of virus-resistant rice (tungro virus), papaya

(PRSV), pepper (CMV) and chilli pepper, while Bangladesh is undertaking research on papaya to develop transgenic cultivars resistant to papaya mosaic virus. India and Pakistan are currently working on the development of virus-resistant rice, cotton, as well as tomato (Pakistan only).

Of all the regions covered by the FAO-BioDeC, Latin America has the highest number of reported research activities regarding the development of transgenic virus-resistant crop varieties. The FAO-BioDeC indicates that nine countries in Latin America are conducting research and/or development on transgenic virus-resistant crop varieties: Mexico, Brazil (13), Cuba, (four) Argentina (two), Colombia (two), Peru, Chile and Costa Rica (two each), and Venezuela (one). No virus-resistant transgenic crop variety has been reported to be commercially released in this region.

In Brazil, field trials are underway for virus-resistant transgenic varieties of sugar cane (SCMV, yellow virus), potato (PVY, PLRV), papaya (PRSV), tobacco (TSWV, PVY), bean (bean golden mosaic virus), *Solanaceae,* and tomato (Gemini and Tospovirus), and laboratory stage experiments are underway on virus-resistant sugar cane. Mexico has virus-resistance trials underway on varieties of papaya (PRSV), potato (PVY and PVX), squash (PAMV, SMV2 and ZaMV), tobacco (TMV), zucchini (PMV, PAMV, SMV2 and ZaMV), melon (CMV), and tomato (CMV). In Cuba, field trials of virus-resistant varieties of papaya (PRSV), and experiments on potato (PLRV), citrus (tristeza virus), and tomato (Gemini virus) are underway. Laboratory stage experiments to develop virus-resistant transgenic varieties are ongoing in Chile for potato and melon, in Costa Rica for rice and maize, in Venezuela for coffee and in Peru for potato and sweet potato.

The large number of research activities reported in the FAO-BioDeC, regarding genetic engineering approaches to generate virus-resistant crop varieties indicates that the basic molecular techniques are well established to develop transgenecassette based approaches for control of most crop viruses. This basic molecular biology capacity is seemingly available in many of the countries covered by the database and the transgene cassettes can often be developed or obtained through collaboration with partner laboratories in other countries with more advanced research capacity. The rapid adoption of this technology within the context of broader agricultural R&D reflects the widespread occurrence of viruses across the regions and crop species, and the difficulty of controlling crop viruses by conventional (non-transgenic) means.

Transgenic Crop Varieties Resistant to Bacterial Diseases

The FAO-BioDeC provides details on the use of genetic engineering for the development of bacterial-resistant transgenic crop varieties. The results suggest a low level of R&D activity in this area in the five regions. Just one transgenic variety with enhanced bacterial resistance is reported as being under field trial - a potato variety with wilt resistance in China. However, laboratory stage research initiatives to develop bacterial resistant crop varieties are reported in six countries: in China; in Thailand; blight resistance in Basmati rice in Pakistan; leaf blight in rice, potato and cabbage in the Republic of Korea ; and also bacterial-resistance work on jute in Bangladesh and banana in Venezuela.

The lower level of activity on resistance to bacterial diseases compared to other diseases may be due to both a lower perception of the importance of bacterial diseases and the number of crops infected by them compared to the incidence of viral diseases, and to more readily

Table 2.

Number of initiatives to develop GMOs with resistance to pests

Trait	*Region*																													
	Africa					*Asia*					*Eastern Europe*					*Latin America*					*Near East*					*Total*				
	C	*T*	*E*	*U*	*N*	*C*	*T*	*E*	*U*	*N*	*C*	*T*	*E*	*U*	*N*	*C*	*T*	*E*	*U*	*N*	*C*	*T*	*E*	*U*	*N*	*C*	*T*	*E*	*U*	*N*
Coleoptera	-	-	-	-	-	-	-	-	-	-	-	-	-	-	-	-	-	-	-	-	-	-	-	-	-	-	-	-	-	-
Lepido-ptera	2	1	1	-	3	3	15	11	6	7	-	-	-	-	-	1	17	5	-	6	-	2	2	-	2	6	35	19	6	18
Other	-	-	2	-	2	-	1	6	8	5	-	-	-	-	-	-	3	10	-	5	-	-	1	-	1	-	4	19	8	13
Total	2	1	3	-	3	3	16	17	14	7	-	-	-	-	-	1	20	15	-	10	-	2	3	-	2	6	39	38	14	22

available alternative technologies to combat bacterial diseases.

Transgenic Crop Varieties Resistant to Fungal Diseases

Some of the most devastating and universal crop diseases are caused by fungal pathogens. For instance, the rust fungi are the most widespread and generally cause the largest crop losses per season. Crop losses can be considerable due to fungal pathogens. For example, the fungal agent of rice blast disease (Magnaporthe grisea) destroys 157 million tonnes of cultivated rice each year, enough rice to feed 60 million people worldwide.

The negative effects of some fungal pathogens can be limited by the use of chemical fungicides. Demand for fungicides amongst farmers is high, indicating that for many farmers there are few available alternatives. The world market for agricultural and non-crop fungicides amounted to an estimated US$6 billion at the end-user level in 1999. The United States, Western Europe and Japan together accounted for 75 percent of the total world market. Small grains constitute the largest market for fungicides worldwide. This sector accounted for an estimated 27 percent of the total world market in 1999, followed closely by tree and vine crops, rice, and vegetables and potatoes. Other crop markets accounted for 17 percent of the world fungicide market, and non-crop markets accounted for 6 percent.

In many countries, fungicides as crop protection products are subject to strict legislative regulation and undergo a rigorous and expensive process of registration for public sale. While fungicides can provide a level of control, this chemical option is often limited for many farmers, particularly in developing countries, by high costs and lack of knowledge about application. In addition, the negative effects of fungicide applications on human health, with special reference to the labourers and the environment can be considerable. There is a need to find more environmentally benign alternatives to fungicides to control fungal diseases of crops.

Genes can be identified that confer resistance to fungal pathogens. For instance, many genes have been found that provide resistance to specific races of each rust pathogen. In many cases, resistance genes are available in the gene pool of cultivated plants and can be transferred to them by cross-breeding programmes. The incorporation of plant-derived resistance genes against fungal pathogens into susceptible varieties could allow development of resistant varieties which can deliver high yields in the absence of fungicide applications.

Actual and potential access by farmers to traditional fungal control measures such as fungicides, and the absence of durable transgenic genetic resistance strategies, may explain why there are few reported efforts to develop transgenic varieties which are resistant to fungal pathogens. In relation to fungi, the current stage of successful research worldwide on identification/isolation of genes conferring durable resistance to fungal diseases of crops, is probably not encouraging developing countries with scarce research resources to embark on transgenic approaches to fungal resistance.

In the African region, only two initiatives for fungal resistant transgenic varieties are reported, a field trial of transgenic strawberry with phytoalexin synthesis genes and laboratory work on transgenic maize for resistance to cob rot (*Stenocarpella maydis*), both in South Africa.. The three initiatives for development of fungal resistant transgenic varieties reported in Eastern Europe were all in Bosnia and Herzegovina where laboratory testing of transgenic potato for resistance to *Fusarium, Verticilium* and *Rhizoctonia* has been initiated. In the Asian region, there is a field trial underway in China for transgenic cotton with resistance to *Verticilium* and *Fusarium*. Other R&D initiatives reported include the involvement of India, Malaysia and Pakistan in research on sheath blight of rice, and of Indonesia on rice blast and leaf rust of coffee.

A few countries in Latin America, mainly Argentina, Brazil and Cuba, are carrying out a number of activities on transgenic resistance to fungi, particularly on tropical fruit trees, with some results already being tested in the field. In this region, most of the activities for transgenic fungal resistance are reported in Cuba, in particular involving field trials of transgenic potato for late blight resistance, and fungal-resistant sugar cane.

Other field trials in the Latin America region for transgenic fungal resistance are reported for maize, sunflower and wheat in Argentina, and tobacco in Mexico. Other crops subject to transgenic R&D for fungal resistance in Cuba are banana, plantain, pineapple, tomato, papaya, citrus and rice. Other countries involved in transgenic fungal resistance research are Argentina on alfalfa, Brazil on rice, barley and cocoa, Chile on grape and apple, Colombia on tree tomato, Peru on potato for late blight resistance and Venezuela on sugar cane.

Resistance to Pests and Herbicides

Resistance to Insect Pests

Insects consume a large share of food and fibre destined for humans (pre- and post-harvest). The worldwide economic damage caused by insect pests to agricultural and horticultural crops and to orchards stands at a hundred billion dollars annually. The strategies to limit the damage from insect pests fall into three categories:

1. Treatment of crops and their pests with chemical insecticides. Drawbacks of the use of chemicals are the development of resistance by the insects to the insecticides and secondly, the toxicity of many pesticides for the environment and human health.
2. Development of insect-resistant crops by hybridisation or by transgenic modification. While transgenic approaches to the development of insect resistance are likely to be the fastest route, or the only means when pest resistance is not available in the genetic pool of the crop plant considered, they are more controversial, as with all genetic modifications of crop plants.
3. Introduction of natural enemies to further biological balance with limited damage. This approach is limited by the availability of natural enemies of the target pest and often risky as it is difficult to predict the outcome of the changes in the agro-ecological niche.

For all these strategies, it is important to gain good knowledge of the pest species that may cause damage at a certain place and time. Quick and reliable identification of the species and monitoring of their geographical distribution and life history is the basis of all effective policies to control insect pests. Worldwide, there are several thousands of insect pest species known to affect a range of crop species, each with their own characteristic damage, distribution and natural enemies. The knowledge and expertise in this scientific field is still limited and often insufficient to develop effective control strategies, especially in tropical or subtropical environments which characterise most developing countries.

Resistant to Lepidopteran Pests

Among the insect pests, Lepidoptera represent a diverse and important group. The FAO-BioDeC indicates that most insect-resistant transgenic crop varieties under R&D are obtained for the control of Lepidoptera (Table 2), predominantly using transgene cassettes including a toxin-producing gene from the soil bacterium *Bacillus thuringiensis* (Bt) see Box 3.

In Asia, *Lepidoptera*-resistant cotton varieties developed using Bt transgenes have been commercially released in China, India and Indonesia and field trials are underway in Thailand. Other field trials underway in China involve local varieties of rice containing Bt transgenes, transgenic cotton expressing Cowpea Trypsin Inhibitor (CPTI), rice transformed for resistance to stem borer and yellow borer, maize for corn borer, cotton for bollworm, and poplar for gypsy moth. Bt tobacco is also in field trials in India and Bt maize in Indonesia.

Laboratory stage research work is ongoing in Asia on potato to generate resistance to tuber moth in India and Indonesia, diamond back moth (cabbage) in India, in addition to generation of resistance to *Lepidopteran* pests of cotton, rice and chickpea in Pakistan. Bt maize has been commercially released in the Philippines, research is underway in Indonesia on transgenic maize for corn borer, soybean for pod borer, rice for stem borer, *Cacao* for fruit borer, sugar cane for stemborer and oil palm for *Setothosea asigna*. Research is also ongoing on transgenic cotton for bollworm resistance in China and Thailand, jute for hairy caterpillar in Bangladesh, soybean in China and Chinese cabbage in the Republic of Korea.

In Latin America, transgenic maize resistant to *Lepidopteran* pests has been released in Argentina, and is in field trials in Brazil and Mexico, and undergoing laboratory testing in Cuba. Other field trials are underway on transgenic soybean in Argentina and Brazil, and cotton in Bolivia, Brazil and Mexico, potato in Mexico and Peru, sugar cane in Brazil and Cuba, tomato in Mexico, sweet potato in Cuba and sunflower in Argentina. Laboratory stage work on transgenic resistance to *Lepidiopteran* pests of rice, coffee and pineapple is ongoing in Cuba and on sugar cane in Brazil.

In the Near East there are field trials of *Lepidoptera*-resistant transgenic varieties of maize and potato in Egypt and rice in the Islamic Republic of Iran , and laboratory work on cotton and maize in Egypt and the Islamic Republic of Iran, respectively. Only four activities to generate *Lepidoptera*-resistant transgenic varieties are reported for Africa. Both *Lepidoptera*-resistant transgenic maize and cotton are under commercial cultivation in South Africa, and laboratory work is underway in Kenya.

Transgenic Bt cotton is reported to be under field trial in Zimbabwe. The rapid adoption of the use of Bt derived transgenes for development of pest-resistant crop varieties, suggests that this approach has generated a new and valuable option for control of some crop pests, which can be applied across a range of agro-environments, crop species and pests.

Resistant to Coleopteran Pests

Beetles and weevils are also important insect pests of crops. A wide range of beetles are of economic importance since they interfere with agricultural and forestry crops, timber products and stored products, etc. However, beetles do not transmit any diseases of humans or livestock. Due to the fact that there are many predators, herbivores and scavengers amongst the beetles, they play an important role in maintaining the ecological balance in natural systems. Furthermore, many host-specific species are used as biocontrol agents of insect pests and noxious weeds.

As an example of *Coleopteran* pests, the African sweet potato weevils (*Cylas puncticollis* and Cylas *brunneus*) are major pests of sweet potato production in sub-Saharan Africa. Sweet potato crop losses due to weevils range from 20100 percent with more severe losses reported in the dry season or during droughts. No genes conferring durable resistance to such weevils have yet been identified within the sweet potato gene pool. The case of sweet potato weevils in Africa is an interesting one regarding decisions on the most effective options for control of the weevils.

There are options for the control of other weevils (e.g. *Cylas formicarius*) used in other regions of the world, which are based on the use of pheromones and bio-insecticides. However, control of African sweet potato weevils has not been successful in Africa and bio-insecticides are generally not available or affordable to African sweet potato farmers. In such a context, the generation of weevil resistant crop varieties using transgenic technology presents a new and potentially valuable option for weevil control. In spite of their agronomic and economic relevance, the FAO-BioDeC (Table 2) suggests that there is no R&D activity to develop transgenic crop varieties which are resistant to Coleoptera pests in any of the regions.

Resistant to Nematode Pests

Nematodes are distributed worldwide and live

saprotrophically or as parasites of plants, animals and humans. While most nematodes in soil are actually beneficial, farmers are most concerned with nematodes that are pathogens of the roots, stems, leaves or seeds of plants. Plant parasitic nematodes are of great economic importance, they are responsible for over US$100 billion in annual crop losses worldwide. They attack a wide variety of plant species, among them many staple crops, vegetables and ornamentals. The root-parasitising nematodes of the genera *Meloidogyne, Heterodera* and *Globodera* are the most important pathogens.

While substantial yield increases have been realised with the discovery and use of nematocides, considerable work remains to be carried out in many aspects of nematode biology and management. On average, only 0.2 percent of the value of crop loss due to nematode damage is invested into nematode research. Furthermore, nematology is a relatively young science and although many crops suffer losses due to nematodes, the number of nematologists working in agriculture is still largely insufficient to properly address this problem. Therefore, much remains to be learned to manage these serious pests.

Due to high crop losses, there is generally a need for the treatment of infested soil with chemical nematocides. Many synthetic chemical nematocides are very unspecific and possess a high general toxicity so that they have been, or will soon be, banned by many governments. Alternative efficient methods to control nematode infestations are rare. Therefore, the search for new nematode control methods is of great economic interest for developing countries.

Strategies deploying crop genetic resistance to nematodes could generate new options for nematode control. The FAO-BioDeC contains no reports in all regions on the use of transgenic technology to develop nematode resistant crop varieties. This was surprising given the high frequency and ubiquity of nematode pest species and the wide range of potential hosting species, across all the regions surveyed. The lack of reports of R&D to develop nematode resistant transgenic crop varieties might be influenced by three factors:

a) the remaining existence on the market of some nematocides that maintain their effectiveness for a considerable number of growing seasons. Many of these nematocides are no longer patent-protected and are available at reasonable cost to some farmers;
b) the limited geographic mobility of the nematode species, which spreads from one field to another very slowly. In many cases, therefore, farmers tend to solve the nematode problem inexpensively by transferring cultivation to other nearby fields; and
c) the slow progress to date in identification of monogenes conferring efficacious resistance to nematodes.

Resistant to Herbicides for Weed Control

Adequate systems for weed control are an essential component of all farming systems. Depending on the crop and weeds present, uncontrolled weeds can reduce yields by over 50 percent, impair crop quality, contaminate the harvest with undesirable weed material, and increase the likelihood that the crop will be attacked by insects or diseases. Weeds compete for nutrients and light, especially during the early stages of crop growth, and for moisture in drought stressed areas, often causing severe yield losses. For example, in labour intensive, small farm operations on the Nigerian savannah, weed-related yield losses from 65 to 92 percent have been recorded.

The parasitic flowering plant known as "witchweed" *(Striga spp.)* remains a major pest of staple crops in sub-Saharan Africa, despite a number of unsuccessful initiatives and considerable research input to find a means of control. The areas infested with parasitic broomrapes (*Orobanche* spp.) and witchweeds are vast and expanding. *Striga* is considered as the greatest single biotic constraint to food production in Africa, where the livelihood of 300 million people is adversely affected. In infested areas, yield losses associated with *Striga* damage are often significant, ranging from 40-100 percent. Crop yields could potentially be doubled if such weeds could be controlled. However, labour intensive weeding is largely ineffective against weeds like *Striga*.

Weeds can be controlled mechanically, chemically or agronomically (e.g. crop rotation). In addition to the undesirable effects of weeds on agricultural production, many weeds can also damage natural areas, alter ecosystem processes and facilitate displacement of native species. Herbicide tolerant varieties provide new options for the control of major weeds which are constraining agricultural production in the regions surveyed. The FAO-BioDeC shows that in relation to herbicide resistant transgenic varieties, the situation is rather different among regions and most of the R&D activity in this area seems to be concentrated in a small number of countries per region. To date, the database suggests that research attention has been focused largely on the development of transgenic crop varieties resistant to glyphosate and glufosinate ammonium. There are also reports of some instances of development of crops resistant to herbicides such as bromoxynil, imidazolinone, bialaphos and isoxazoles.

Glyphosate resistance has been most widely used in Latin America, where Roundup-Ready™ (RR) transgenic soybean has been cultivated in Argentina and Uruguay. According to other reports, the same two countries have released two more herbicide-resistant soybean varieties, however, without specification of the kind of herbicide resistance. A glufosinate-resistant transgenic maize variety has also been released in Argentina. Brazil has recently accepted the RR soybean and is now cultivating it on a large scale. Thirty-four herbicide tolerant transgenic varieties are being field tested in Latin American countries, particularly in Argentina, Brazil and Mexico. In this regard, field trials have been performed for transgenic cotton, alfalfa, maize, sunflower, sugar beet and wheat in Argentina, maize, cotton, sugar cane and eucalyptus in Brazil, and cotton, maize and soybean in Mexico.

In addition to glyphosate, glufosinate resistant varieties are important in Latin America, where 14 field trials have been carried out in the region on transgenic soybean in Argentina, Brazil and Mexico, sugar cane in Brazil and Cuba, maize in Brazil and Mexico, wheat in Argentina and Mexico, sugarbeet in Argentina, rice in Brazil and potato in Cuba, which also has extensive laboratory input into rice, banana, plantain, coffee and pineapple. However, it must be noted that the number of laboratory research activities in the area of glufosinate resistance is much lower, and all such research activities are concentrated in Argentina (barley and sugar cane), Cuba, (rice, banana and plantain, coffee and pineapple) and Venezuela.

In Eastern Europe, field trials on glyphosate and glufosinate tolerant maize are ongoing in Serbia and

Montenegro, whereas Bulgaria is commercially cultivating unspecified herbicide tolerant maize. In the Near East, the only research reported is on transgenic canola in the Islamic Republic of Iran. In Asia, Indonesia is conducting herbicide tolerance field tests on imported GM maize, cotton and soybean, and China on soybean. China is conducting research on herbicide tolerant rice, Pakistan on wheat and the Republic of Korea on cabbage, chinese cabbage and potato, though no details are given.

In Africa, the only GM herbicide tolerance research reported in the FAO-BioDeC is taking place in South Africa, with field trials of 11 transgenic varieties resistant to different herbicides. The crops involved included canola, cotton, *Eucalyptus*, lucerne, maize, soybean, sugar cane and strawberry.

In all regions, the higher number of field trials when compared to laboratory stage research in this area, suggests that most of the varieties undergoing field trials are likely to have been generated outside of the regions or result from crosses with local varieties. In addition, the relatively high number of laboratory research in this area compared to e.g bacteria resistance may also reflect the comparative technical simplicity for many species of generating herbicide tolerant crops using transgene cassettes obtained from research institutions or companies with more advanced research capacity.

The limiting factors to R&D to generate herbicide tolerant transgenic crops are the number and type of herbicides and transgene resistance cassettes available, the ease with which the crop species or variety can be transformed, the registration requirements both for the herbicide and the transgenic crop variety, and the level of access of farmers and researchers to proprietary herbicides and transgene resistance cassettes. Yet, the early experiences with herbicide resistant maize to control

Table 3

Number of initiatives to develop GMOs with resistance to herbicides

Trait	Region: Africa					Asia					Eastern Europe					Latin America					Near East					Total				
	C	T	E	U	N	C	T	E	U	N	C	T	E	U	N	C	T	E	U	N	C	T	E	U	N	C	T	E	U	N
Glufosinate	-	3	-	-	1	-	-	-	-	-	-	-	1	-	1	1	14	5	-	4	-	-	-	-	-	1	17	6	-	6
Glyphosate	-	4	-	-	1	-	4	-	-	2	-	1	-	-	1	3	14	-	-	4	-	-	1	-	1	3	23	1	-	6
Phosphinotricin	-	2	-	-	1	-	-	-	-	-	-	-	-	-	-	-	-	-	-	-	-	-	-	-	-	-	2	-	-	1
Other	-	2	1	-	1	-	1	-	6	3	1	-	-	-	1	-	6	6	-	4	-	-	-	-	-	1	9	7	6	8
Total	-	11	1	-	1	-	5	-	6	4	1	1	1	-	2	4	34	11	-	7	-	-	1	-	1	5	51	14	6	15

Striga infestations in Africa suggest that novel models to disseminate herbicide tolerant transgenic crops could benefit even resource-poor farmers.

Tolerant to Abiotic Stresses Resistance

Abiotic stresses continue to limit crop productivity in every season and in every agro-ecosystem worldwide. Among the abiotic stresses affecting crop production in developing countries, drought and low soil fertility are the most significant. Plants vary tremendously in their ability to withstand abiotic stresses, both between species and within populations of a single species, but the nature of abiotic stress tolerance is not well characterised. Understanding the mechanisms of abiotic stress tolerance will have a significant impact on crop productivity. For instance, crop loss to drought in the tropics alone is thought to exceed 20 million tonnes of grain equivalent per year, or around 17 percent of well-watered production, reaching up to 60 percent in severely affected regions such as southern Africa in 1991-92. The development of crop varieties which can tolerate abiotic stresses would be of major benefit to agriculture in cropping regions where abiotic stresses are a chronic problem.

As shown in Table 4, no transgenic crop variety tolerant to abiotic stress has so far been reported to be released for cultivation in any of the regions covered by the database. However, 7 transgenic varieties exhibiting tolerance to a range of abiotic stresses underwent field testing in Bolivia, China, Egypt, India and Thailand. The number of research initiatives at the laboratory stage in this area totals 27.

Most of the R&D activities on development of abiotic stress tolerant crops are being carried out in six countries of the Asian region, namely Bangladesh, China, India, Indonesia, Pakistan and Thailand. China is relatively active in this area, and has reported preliminary successes with rice, maize and sorghum tolerant to high salt concentrations. Transgenic studies on salt resistant crop development are also being undertaken on rice in Bangladesh, Brazil, India and Pakistan, and on tobacco in Argentina. Transgenic research approaches to obtaining aluminium-resistant varieties are underway for wheat in Mexico and sugarbeet in China.

Despite the effects of drought on crop production, very little transgenic research on drought resistance is reportedly being carried out in the five regions. Drought-tolerant rice is currently being field tested in Thailand. The rest of the activities in this area are mainly at the laboratory stage, with work on sugar cane (Indonesia), rice (China and Indonesia), and groundnut (South Africa). Overall, the extent of R&D reportedly devoted to abiotic stress tolerance is insufficient when compared to the well known needs for abiotic stress tolerant varieties in the regions surveyed.

For instance, vast areas of soils containing an excess of heavy metals are present in Brazil and Africa. Also, a steadily increasing acreage of agricultural land in Asia and elsewhere is becoming agriculturally sterile because of salinity brought about by poorly managed irrigation practices. The major limitation is the complexity of tolerance to abiotic stresses that is normally dependent on a number of physiological traits, each under multigenic control. The increased wealth of knowledge that is being acquired by means of genomics, mainly functional genomics, and other molecular biology studies, will certainly contribute to the development of tolerant genotypes.

Modified Quality Traits Resistance

The nutritional and economic value of many grain crops depends on the quantity and types of compounds that accumulate during seed development. Plant breeders have long strived to develop new crop varieties with enhanced quality trait profiles. The results shown in Table 5 describe the different situations, regarding the development of transgenic crops with enhanced quality traits in each region. No research or field testing activity is reportedly taking place in the African or Near Eastern regions, despite well-known dietary deficiencies prevailing in both regions. However, this situation is not paralleled in Asia, where a total of 34 activities on this topic are reported in eight countries or in Latin America where 31 reported initiatives in eight countries. In Latin American countries, there are 12 research activities reported and 19 transgenics under field testing, mostly in Argentina and Mexico.

A wide range of characteristics can fall under the topic of 'quality trait' and the use of this loose term in a survey can result in different categorisations by different respondents. For instance, the reports from Latin America indicate that there are field trials of varieties with enhanced quality traits underway on rice (Sucrose-phosphate synthase) and other trials for unspecified characteristics are underway on soybean in Argentina, and canola and flax in Mexico. Laboratory studies cover alfalfa (veterinary edible vaccines), *Paspalum dilatum* (foddering quality), *Triticale* (biomass production) in Argentina, eucalyptus (reduction of lignin content) and *Psychotria* spp. (improved alkaloid production) in Brazil, sugar cane (altered lignin content and high quality sugar) in Cuba, *Echinaceae*, *Psychotria* and *Tagetes* (alkaloid production) in Costa Rica, and improvement of flour

Table 4

Number of initiatives to develop GMOs with resistance to abiotic stresses

Trait	*Region*																													
	Africa					*Asia*					*Eastern Europe*					*Latin America*					*Near East*					*Total*				
	C	T	E	U	N	C	T	E	U	N	C	T	E	U	N	C	T	E	U	N	C	T	E	U	N	C	T	E	U	N
Frost	-	-	-	-	-	-	1	-	1	2	-	-	-	-	-	-	1	-	-	1	-	-	-	-	-	-	2	-	1	3
Salt	-	-	-	-	-	-	1	4	2	5	-	-	-	-	-	-	-	2	-	2	-	1	1	-	1	-	2	7	2	8
Heat	-	-	-	-	-	-	-	-	-	-	-	-	2	-	1	-	-	-	-	-	-	-	2	-	1	-	-	4		2
Drought	-	-	1	-	1	-	2	1	2	3	-	-	1	-	1	-	-	-	-	-	-	-	1	-	1	-	2	4	2	6
Other	-	-	1	-	1	-	1	2	1	2	-	-	-	-	-	-	-	7	-	5	-	-	2	-	1	-	1	12	1	9
Total	-	-	2	-	2	-	5	7	6	6	-	-	3	-	1	-	1	9	-	5	-	1	6	-	1	-	1	27	6	15

quality and reduction of natural toxicants in potato in Peru.

Quality traits also exist in the floriculture industry. In this regard, a petunia variety with altered flower colour is now in commercial production in China. Micronutrient malnutrition, especially lack of iron, zinc and vitamin A, currently afflicts more than half the world's population. Called "micronutrients" because they are needed in only miniscule amounts, these substances enable the body to produce enzymes, hormones and other substances essential for proper growth and development. Tiny though the amounts are, the consequences of their absence are severe.

Iodine, vitamin A and iron are most important in global public health terms. Their lack represents a major threat to the health and development of populations the world over, particularly to pre-school children and pregnant women in low-income countries. Enhancing the micronutrient (vitamin and mineral content) status of staple crops is considered to be one approach where crop biotechnology could generate crop varieties that could be used to strengthen food security and prevent malnutrition.

Protein Content and Amino-acid Profiles

Protein-energy malnutrition (PEM) is a most lethal form of malnutrition. Children are its most visible victims. Malnutrition contributes to at least half of the 10.9 million child deaths each year. PEM disorders result from a lack of protein and carbohydrates. Cereals and root crops such as cassava form a substantial portion of developing country staple diets, thus improving their protein content and quality will go a long way to curbing the devastating effects of PEM. Seed proteins of many major crops often do not contain sufficient quantities of amino acids essential in the diet of humans and other monogastric animals. As cereals are usually deficient in lysine and tryptophan, and legumes are deficient in the sulphur amino acids methionine and cystine, a mixture of cereals and legumes is used to provide a balance of amino acids in the diet. Breeders at CIMMYT (Mexico) have managed with conventional breeding to develop quality protein maize (QPM) varieties which have enhanced levels of the two 'essential' amino acids, lysine and tryptophan, in the endosperm protein. These new varieties look and taste like normal maize but the nutritive value of their protein is nearly equivalent to cow's milk. Modifying the genes that encode seed proteins by genetic engineering is one additional option to address the problem of nutritional quality in some staple crop varieties.

The FAO-BioDeC does not reveal any significant level of research on metabolic engineering of protein content nor quality in crop varieties in the vast majority of countries surveyed. The only research reported is on field trials for high lysine maize in China. Argentina is also reported to have conducted field trials for unspecified traits in maize and soybeans as well as wheat with high molecular weight glutein.

Vitamin Profiles

Vitamins are essential components of the human diet and dietary deficiencies in some vitamins can have tragic effects. For instance, vitamin A deficiency (VAD) is the leading cause of preventable blindness in children and raises the risk of disease and death from severe infections. In pregnant women, VAD causes night blindness and may increase the risk of maternal mortality. It is a public health problem in 118 countries, especially in Africa and South-East Asia, once again hitting hardest young children and pregnant women in low-income countries.

Crucial for maternal and child survival, supplying adequate vitamin A in high-risk areas can significantly reduce mortality.

The arsenal of nutritional approaches to combat VAD includes a combination of breastfeeding and vitamin A supplementation, coupled with enduring solutions, such as the promotion of vitamin A-rich diets and food fortification. Among these approaches, biotechnology research has contributed the additional option of increasing the vitamin A content of staple foods which are typically low in vitamin A content. The proof of concept of this approach has been the highly publicised high-beta carotene (pro-vitamin A) rice transgenics recently developed, and currently under dissemination to many NARS, under the name of 'golden rice' and 'golden mustard' in India. Yet, the FAO-BioDeC indicates that there is little R&D underway on enhancement of vitamin content in crop varieties in the regions surveyed.

Source of Minerals

Most staple crops are not considered an important source of minerals in the human or domestic animal diet. Yet, because of the high level of consumption of staples, small increases in the mineral concentration of staples could have a significant effect on human nutrition and health. Iron deficiency anaemia afflicts an estimated 1.5 billion people in developing countries, most of them women, reducing mental ability, creating severe complications at childbirth, and lowering physical capacity. Zinc deficiency, though less well understood, is also known to be widespread in the tropics and is a major threat to children's growth and health. The nutritional quality of staple crops in terms of mineral content can be improved by both conventional breeding and/or biotechnology. Currently the FAO-BioDeC does not report any biotechnology activity targeted towards the improvement of crop mineral profiles in developing countries.

Oil Composition

It seems from the countries covered by the FAO-BioDeC that the modification of plant oil and wax composition has received little attention. Among the instances of R&D in this area are ongoing field trials in Argentina on maize and soybean, and in Mexico on canola expressing high levels of lauric acid. In addition, there is ongoing laboratory work on oil palm with low saturated fatty acids in Indonesia and Malaysia and high lauric acid content in coconut in the Philippines. Furthermore, in Malaysia there is development of oil palm with special oils for the production of biodegradable plastics.

Plant Growth Traits

The genetic manipulation of the metabolic routes leading to plant hormone (e.g. ethylene) synthesis and degradation can induce modifications of plant organ maturation. Such control of maturation can allow the production of fruits showing resistance to postmaturation deterioration, resulting in the ability to be transported without refrigeration, extended shelf life and improved quality. For many developing countries with limited refrigeration capacity and transport infrastructures, this research avenue is extremely promising, insofar as it improves flexibility of transport of otherwise perishable products to distant markets and on poor routes.

Most of the research reported on alterations in plant growth has concentrated on delayed fruit ripening. In Latin America, Mexico reports a large number of field trials for altered growth in varieties of crops such as tomato, chilli pepper, banana, melon, papaya and

Table 5.
Number of initiatives to develop GMOs with improved quality traits

Trait	Africa					Asia					Eastern Europe					Region Latin America					Near East					Total				
	C	T	E	U	N	C	T	E	U	N	C	T	E	U	N	C	T	E	U	N	C	T	E	U	N	C	T	E	U	N
Vitamin content -	-	-	-	-	-	-	-	2	2	-	-	-	-	-	-	-	-	-	-	-	-	-	-	-	-	-	-	2	2	
Oil composition-	-	-	-	-	-	-	-	-	-	-	-	-	-	-	-	3	-	-	3	-	-	-	-	-	-	3			3	
Protein content -	-	-	-	-	-	1	1	-	2	-	-	-	-	-	-	-	-	-	-	-	-	-	-	-	-	1	1		2	
AA composition-	-	-	-	-	-	-	-	-	-	-	-	-	-	-	-	3	-	-	3	-	-	-	-	-	-	3			3	
Starch composition-	-	-	-	-	-	-	2	-	2	-	-	1	-	1	-	-	-	-	-	-	-	-	-	-	-		3		3	
Altered growth traits -	-	-	-	-	1	2	5	-	4	-	-	-	-	-	-	6	1	-	2	-	-	1	-	1	1	8	7		6	
Other -	-	-	-	-	1	-	19	2	6	-	-	9	-	5	-	7	11	-	7	-	-	-	-	-	1	7	39	2	18	
Total -	-	-	-	-	2	3	27	4	8	-	-	10	-	5	-	19	12	-	8	-	-	1	-	1	2	22	50	4	22	

pineapple. In this area, Chile has a laboratory research initiative on stone fruits. In Asia, papaya varieties with altered growth traits are under laboratory study in China, Indonesia and the Philippines. In the latter, mango is also being researched, as is tobacco for delayed leaf senescence. Other growth phase changes receiving attention are short-stature and profuse tillering in basmati rice in Pakistan. In Eastern Europe, laboratory work in this area is ongoing in Bosnia and Herzegovina in potato for higher cytokinin levels.

Transgenes—Generating Crop Varieties with Multiple Transgenic Traits

The first generation of commercially available transgenic crop varieties has typically included cases where one transgene has been used to add an enhanced characteristic to a particular crop variety. Yet, there are a multitude of characteristics which are assessed by farmers to determine whether a variety is suitable or unsuitable.

The production of crops with multiple genetically engineered traits may seem a logical step to follow and, as more locally adapted transgenic varieties become available and accepted on the market, these will become the source material into which novel genes will be incorporated. However, some problems can arise if too many transgenes are incorporated into a single variety, and such effects may be cumulative over successive generations.

To overcome such problems, novel approaches to simplify and improve the process of introducing multiple transgenes into crop varieties are under development. Where data on the use of multiple transgenes to develop single crop varieties are present in the FAO-BioDeC, the most commonly used dual-transgene combination is herbicide resistance combined with insect resistance.

In Latin America, Argentina has released transgenic maize with *Lepidoptera* resistance and glufosinate tolerance and there are 11 field trials reported in the region on GM crops. The absence of laboratory work suggests that this material is a technology import. Transgenic traits of *Lepidoptera* resistance with glyphosate tolerance are undergoing field trials in alfalfa, maize, soybean and sunflower in Argentina, and in cotton in Argentina, Brazil and Mexico, where maize with *Coloeptera, Lepidoptera* resistance and glyphosate tolerance is also being field tested. *Lepidoptera* resistance combined with glufosinate tolerance is undergoing field testing in maize in Argentina.

In South Africa, two field trials tested insect resistance combined with bromoxynil or glyphosate tolerance, and maize with potential insect resistance and glyphosate tolerance was field tested. Other traits that are combined with herbicide tolerance are fungal resistance with glufosinate tolerance in wheat in Argentina, and unspecified pathogen resistance with glyphosate tolerance in Mexico. In Argentina, *Coleoptera* and PVY resistant potato is undergoing field trials. In Asia, two field trials are ongoing in China to test transgenic rice for combined blight and RDV resistance and combined bacterial wilt and PVY resistance in potato, as well as salt and herbicide tolerance in potato.

In the Philippines, experiments to combine fungal, insect and bacterial resistance with salt tolerance in rice are underway. Papaya is receiving attention for combined ringspot virus resistance and delayed ripening in the Philippines, and similar work to combine virus resistance with extended shelf life is reported in Malaysia. Golden rice is a multi-trait transgenic plant since it was made by the transfer of three different transgenes for the synthesis

- β-carotene taken from daffodil (*Narcissus pseudonarcissus*) and bacterium *Erwinia uredovora.* Metabolic pathway engineering is technically complex and challenging, and according to the FAO-BioDeC it has not yet been carried out in any crop within a developing country setting.

5

Traditional Technologies for Agricultural Production

More than 60 animal species are directly useful to humans, but most attention tends to be given to cattle, buffaloes, sheep, goats, pigs, horses, donkeys and poultry. There are also more unconventional animals such as llamas, yaks, guinea fowl, ducks, bees and pigeons that can adapt to many conditions. Often these unconventional animal species consist of small animals that have the advantage of fast reproduction, i.e. a herd or flock of these species is quickly replaced after a calamity such as drought, a flood or disease outbreak. Large unconventional livestock such as camels, llamas, alpacas, yaks, bantengs and deer are adapted to specific ecological niches, often in mixed systems.

Animal health is essential in mixed farming systems, but veterinary health care knows failures as well as successes. The major successes are those of vaccination campaigns; the major failures occur where the veterinary services are largely unable to reach small farmers with their individual problems. Veterinary care in terms of prevention can help to improve livestock production. However, in traditional livestock systems the cattle owner mainly tries to avoid major damage at mini mum cost and without depending on expensive and possibly erratic services from outside.

Part of the health care therefore lies with local technicians using indigenous veterinary medicine, and learning to cope with disease—by spreading risks, by using animals tolerant to local diseases, and by running a low-cost operation. Many diseases occur naturally and regularly depending on local ecological and sociocultural conditions, but changes in the world have also had their impact on local systems with consequences for animal health:

— War has forced people to leave their villages. By taking their animals with them, some diseases have spread to neighbouring areas or countries. For example, Mozambican refugees from the north spread African swine fever to peri-urban settlements around Maputo in the south.

— Population growth increases the pressure on land. Cattle and small ruminants have to move further from the village to less favourable pastures, which causes stress, and in addition there may be more contact with herds from other villages.

— Commercial movement of animals has led to the spread of disease. For example, rinderpest was introduced into Africa by a shipment of infected cattle from Asia in 1889 to feed Italian soldiers. It was catastrophic for cattle and wildlife. To replace these losses, in around 1903, cattle from Tanganyika were moved to Zimbabwe and this shipment spread theileriosis over southern Africa.

The actions that can be taken to cope with the effect of these changes are many, though not all problems can be solved on a small scale. One technological and futuristic approach lies in better mapping and prediction of animal diseases. Another lies in the continuation or modification of vaccination campaigns. Still another approach lies in the use of barefoot doctors and in recognition of the benefits of indigenous medicine. Some of these approaches relevant to mixed farming systems are given below.

Traditional and Modern Treatments of Vector-borne Diseases

Humid tropical regions harbour a range of endemic infectious animal diseases. Many of these diseases are caused by micro-organisms that hide for part of their lives in a tick, a mosquito or a fly. Disease pressure is related to the density and the species of the vector. Cattle reared in infested areas for thousands of years have acquired a certain natural resistance but animals may still suffer and die when the infection pressure is high, or when other infections burden the immune system. One way to overcome or prevent these vector-borne diseases is to use insecticides in so-called cattle dips. This allows farmers to maintain animals with higher production levels because natural resistance tends to be at the expense of milk and meat output. However, breeds with natural resistance can be reared with a minimum of expense and they will continue to be important in sustainable livestock production in Africa.

The main disease vectors in cattle are the tsetse fly, various tick species and some mosquitoes. Over one-third of the African continent is infested by the tsetse fly, which transmits the micro-organism causing trypanosomiasis. Local cattle and goats have more genetic resistance to these micro-organisms than exotic breeds with higher milk or meat yields. The best-known tick-borne diseases of cattle are theileriosis, anaplasmosis, babesiosis and cowdriosis. African swine fever is a tick-borne disease of pigs to which exotic breeds are highly susceptible.

Vector-borne micro-organisms are hard to control but scientists and farmers continue to try with varying levels of success. A traditional way to protect livestock is to avoid areas of high risk in certain periods. For example, Fulani cattle keepers in the subhumid zone of West Africa avoid infested grazing areas, and they minimise the time spent at watering points where vectors are most likely to occur. In the wet season, grazing is delayed until late in the morning, as worm infestation on grass is highest early in the morning.

Fires are built next to cattle pens to keep away biting insects at night and, when outbreaks do occur, the livestock owners avoid infected areas. Because of increased population pressure, however, it is increasingly difficult for nomadic cattle owners to avoid unfavourable areas, even by trekking long distances. The most economic approaches from formal research combine control and protection measures in a ratio that depends on socio- economic and epidemiological conditions.

Veterinary and animal health research uses genetic approaches in cattle, sheep and goats against vector-borne diseases. They work primarily on the genetic resistance of cattle breeds to the micro-organisms and to tick vectors, on genetic resistance of goat breeds to trypanosomiasis and on cattle breeds with natural resistance to ticks. In addition, genetic engineering of breeds with genes that favour resistance against diseases is a relatively new avenue in the battle against animal diseases and not yet proven to be effective.

Another approach aims to reduce the vector burden, as has been done for decades, for example by the dipping of cattle against ticks. This is successful in the short term but it does not have a lasting impact on the tick population, nor on the disease. In addition, ticks develop resistance to acaricides. New acaricides will have to be developed continuously, to stay ahead of the diminishing sensitivity of the ticks. Resistance continues to develop against these products as well, i.e. new products need to be developed continuously to stay ahead. The FAO strategy on ticks and tick-borne diseases programmes aims at promoting integrated tick and tick-borne disease control methods that include immunisation when applicable and increasing awareness about resistance of ticks to acaricides.

The development of vaccines against tick-borne diseases and trypanosomiasis is a major line of research. Important progress has been made in understanding the way in which micro-organisms attack the host. The vaccines so far developed all have disadvantages. East Coast fever vaccine, for instance, is expensive and exists off live, virulent strains. Animals vaccinated for East Coast fever may infect ticks long after vaccination and contribute to new outbreaks in subse quent years. Immunisation against some tick species has been tried and may have some possibilities. When this approach has effect, the tick population could be suppressed without damage to the environment.

The interest in ethnoveterinary practices is growing. Many of these practices offer viable alternatives to conventional western-style veterinary medicine especially where the latter is unavailable, unaffordable or inappropriate. Ethnoveterinary medicine can provide low-cost health care for simple animal health issues though it tends to be ineffective against infectious diseases. Ethnoveterinary remedies are often based on knowledge and tradition from folk medicine for human use, such as in India on the *ayurveda*.

Most of the plants used are easily available but non-plant substances are also used. For example, warm stout is given to animals after they have given birth to help remove the afterbirth and cobwebs are used on cuts to help stop the bleeding. Some of the plants used are multipurpose such as guava (*Psidium guajava*), bamboo (*Bambusa vulgaris*), rice (*Oryza sativa*), turmeric (*Curcuma longa*), aloe (*Aloe vera*), banana (*Musa spp.*) and *Kalanchoe pinnata*. These plants are either already found on farms or they can easily be grown. Many of these plants also have a food value.

For example, an excess of green bananas can be ground, boiled and fed to stock as a source of carbohydrates and iron. Guava fruits and leaves contain useful vitamins. *Cymbopogon citratus* and *Ocimum gratissimum* can be used to make delicious teas. Medicinal plants to treat ruminants are used mainly for internal parasites, internal and external injuries and pregnancy-related conditions. Farmers usually boil the plants to make a decoction. Other plants are administered as teas, in which water is boiled and thrown on to the fresh leaves, which are left to steep and then administered once or over a period of days. Bamboo joints, thin-necked bottles or other appropriate instruments are used to drench the animals.

As with any technology, care has to be taken in the use of indigenous medicine and application of knowledge. However, more attention to the potential of these approaches is likely to unlock a vast area of useful knowledge for conditions where modern medicine is out of reach.

Housing and Management

Domesticated animals live, by definition, in and around the homesteads of people. Animals are either kept literally in the house, or in a pen or stable; they can be herded at shorter or larger distances from the home or they are only occasionally rounded up in a cattle crush or corral. Animals are not kept in sheds in conditions of EXPAGR, where land is abundant. Occasionally they are kept in corrals that are used for deticking of the animals, for marking, for vaccination or for castration.

Housing for animals demands investment and farmers must carefully consider whether it brings sufficient advantages to warrant such an investment. Many intermediate solutions exist, from grazing behind fences, to tethering cattle, to zero grazing where the feed is brought to the animals and where the dung is collected. Animals may be kept in stables year-round or only seasonally. During wet growing seasons animals tend to be kept on the homestead whereas afterwards they are allowed to graze on the harvested fields and in the bush.

In semi-zero grazing systems, animals are kept in a stable or fenced enclosure for part of the day and particularly during the night, where they may be given some cut fodder. For the remainder of the day, they are allowed to graze. The benefits, i.e. the reasons for housing and management, can vary widely between systems but they include aspects such as:

- Protection against climatic conditions
- Excreta management
- Disease control
- Prevention of theft or damage by predators
- Saving of labour

— Control of product quality
— Prevention of damage to crops

Animal housing in the warm tropics requires a shed design that keeps out rain and solar radiation, while allowing the free flow of air. Ventilation helps to cool the animals by allowing evaporation of water, thus cooling the air; by keeping the sun out, the place is also kept cool. In the hot tropics walls can be absent, although thick brick, stone or mud walls can help to lower temperature fluctuations between day and night. Roof overhang is important to keep out the sun and rainstorms when walls are absent. Trees provide shade and fresh air around the stable. They can be used to store feed and roofing material should reflect the solar radiation and allow ventilation while preventing draught. To deflect solar radiation the roof can be painted white, made of reflecting materials, of tiles and/or plant materials that insulate (such as straw, grass or palm tree leaves).

Feed provides the energy and minerals that allow animals to stay alive, to grow, to produce and reproduce, etc. It can come as straw, grass, tree leaves, grains, tubers, insects, etc., depending on the agroclimatological conditions and on the mode of farming. Livestock in mixed systems of the EXPAGR mode depend mainly on grazing on wastelands, fallowed croplands or distant grazing areas. As the systems move into the LEIA mode there is a greater need to use crop residues or industrial wastes. As systems move further into NCA there is a particular opportunity to use leaves from leguminous crops, or biomass from crops that are grown between the main crops. Improved fallows, so-called leys, can use crops such as legumes or crucifereae and even fodder grains, while living fences provide timber, fodder and posts.

In HEIA systems it is common to grow special fodders or to buy feed from outside; mixed farms where animals and crops exchange resources "on the farm" are rare in HEIA. Crop residues are relatively unimportant where there is plenty of land or other resources such as in EXPAGR and HEIA. However, use of crop residues for feed becomes vital in situations of LEIA and NCA where land becomes scarce. The straws from cereals, sometimes coarse such as maize, sorghum and pearl millet, and sometimes fine and slender such as wheat, oat or barley stover, thus form a major feed resource under many LEIA conditions.

Unfortunately, the feeding value of such straw and stover is low. Cattle and buffaloes can eat little of them and the concentration of nutrients is low. As feeds become scarcer and of less value, it is better to give one feed in a less wasteful way and/or introduce methods to overcome the nutritional limitations, e.g. through treatment with chemical or physical methods, through choice of other varieties and planting methods, etc.

Feeding of Fibrous Crop Residues

Production of crops such as cereals and oilseeds yields two kinds of by-products that can be used as animal feed—the highly valuable grain and oilseed residues and the poor quality straws and stovers. The green fodders from legumes and specialised grass production that are part of the crop rotation occupy an intermediate position in terms of feed quality (Table 1). Animals at low levels of production of milk, meat, draught, etc. do not generally require much supplementation to meet their nutrient requirements. They obtain sufficient nutrients by grazing on roadsides, or on fallow lands, or by consuming crop residues and kitchen waste at home.

During periods of scarcity they lose some bodyweight, which they can regain during the more favourable season. Low-quality feed does not provide sufficient nutrients to maintain the bodyweight of ruminant animals but it may be sufficient for survival. To achieve any production of milk, meat or draught over long periods, low-quality fodders must be supplemented with better feeds and/or they must be treated with chemicals or other methods. Farmers commonly supplement low-quality fodder with concentrate feed ingredients such as brans, cakes or mixed commercial feeds. Treatments are less commonly used and they are useful only under specific conditions. Different treatment methods and supplementation are summarised in Table 2.

Table 1. Classification of crop residues according to crude protein content (CP), energy content (TDN) and CP:TDN ratio

Crop residue type	*CP (%)*	*TDN (%)*	*CP:TDN*
Category I: good quality			
Oilseed cake	28	70	0.40
Concentrate feed	15	65	0.23
Legume tree leaf	24	60	0.40
Category II: medium quality			
Medium quality grass	12	60	0.20
Rice bran	11	55	0.20
Mature grasses	10	55	0.18
Category III: poor quality			
Maize straw	6	50	0.12
Rice straw	4	45	0.09

Special feeds to supplement the lack of nutrients in straw-based rations can be used in several ways, called catalytic, strategic or substitutional supplementation. The latter is only applied in industrial systems where cheap concentrates/brans/cakes are available. Under these conditions straw is only fed as a source of fibre to aid digestion. The aim in these systems is to feed as much concentrate as possible, i.e. to substitute the roughage.

Table 2. Feeding systems based on the use of fibrous crop residues

Feeding system	*Description*
Emergency and survival feeding	The use of any type of feed to achieve survival of the herd or animal, if necessary at the expense of liveweight and/or re)production.
Catalytic supplementation	The use of small quantities of good-quality feed to improve digestion and intake of a basal ration of straw or mature grass.
Substitutional supplementation	The use of large quantities of supple--ment to supply sufficient nutrients for a desired level of animal output, if necessary at the expense of straw/grass intake, i.e. supplement substitutes the basal ration.
Straw treatment	Use of physical or chemical methods to increase straw feeding value or consumption
Chopping and soaking	Chopping implies the reduction of feed particle size, commonly at the size of a few centimetres or more, mostly to avoid waste of feed. Done alone or in combination with soaking.
Selective consumption	Farmers and/or animals can select the good part of the feed, leaving the residue for animals of lower output or for other uses than feed.
Stripping	The use of leaves before they mature on plants for animal feed; mostly coarse grains such as maize and millets.
Thinning	The use of purposely dense sown plants for animal feed that are thinned and fed to the animals as the crop matures.

Contd....

Contd....

Use of variability	The term "use of variability" implies the use of differences in straw quality and quantity due to management, environment and genetic factors.
Adjusted cropping	A variation on the theme of variability, i.e. crop choice is at least partly based on the nutrient requirements of the animals; or animals and crops are mutually adjusted, e.g. in the Flemish/Norfolk systems.

Catalytic Supplementation

Supplementation of low-quality fodder with nutrients such as nitrogen and/or phosphorus is called catalytic supplementation if it is done to achieve improved rumen function. Small amounts of nutrients can result in increased intake and digestibility of straws. The supplement is said to have a positive associative or "catalytic" effect on digestion. The approach aims to increase the roughage utilisation by using a minimum quantity of supplement.

Obviously, this is attractive where low-quality roughage is cheap and where supplement is expensive, in combination with production objectives that aim at near maintenance growth. The most common example of catalytic feeding is the use of lickblocks or other systems in which cattle can lick a mixture of urea and molasses. Lickblocks often contain urea as a supplementary element, as it needs to be fed slowly and distributed evenly. The block is a compact brick, placed before the animal to lick. This helps to meet the nitrogen requirement of rumen microflora, resulting in enhanced rumen activity and increased degradability of crop residues. Feeding of some kitchen waste or green leaves may achieve the same objective of maintenance.

Strategic Supplementation and Selective Consumption

Another approach for efficient use of a limited stock of supplements is to use them only for certain animals: strategic supplementation. It is applied where better feeds are given to reproductive and young valuable animals, while adults or less valuable animals are left to lose weight. A typical but disguised form of strategic supplementation is the use of selective consumption. In this case the animals are offered a large amount of feed from which they can select the better parts such as the leaves while rejecting the rest such as the stems.

The inferior feed refusals are then fed to less productive or idling animals or used for bedding. The final residue may be mixed with dung for composting or dung cakes. Strategic supplementation is applied frequently on farms when not enough better-quality fodder or feed are available for all the animals. Lactating cows, sick and weak animals or draught animals receive extra feed to perform their duties, to get well or do the fieldwork, respectively.

Treatment of Straw

A range of treatment methods is available to increase digestibility and/or intake of straw. Simple methods include chopping or soaking of straw and more complicated ones involve steam treatment, e.g. for the treatment of sugar-cane bagasse. The most practical approach is based on the use of urea. Urea can be sprayed over the straw in a ratio of 2 percent with a 1:1 water:straw ratio, a form of catalytic supplementation. More relevant is the treatment of straw with urea where 4 kg of urea are spread with 50-100 litres of water on 100 kg of straw. The mix is kept in a heap for one to three weeks after which it can be used as feed with or without concentrate supplements.

The treatment process increases the availability of energy from the fibres in the straw, apart from providing nitrogen for better rumen function. Experience suggests that the technology is most likely to work in the following situations:

- when plenty of dry straw is available, free from fungal contamination;
- when farmers have slender straws from rice, wheat and barley rather than coarse straws from maize, sorghum or millets;
- when straw is cheap compared to other feeds and where there is a shortage of grasses or other green feed;
- when water is freely and conveniently available;
- when the price of urea is not prohibitive and where the cost of polythene covering materials is low;
- when labour availability is good, though small stacks do not require such high labour inputs at one time as the large stacks;
- when the animals are medium producers (milk or meat);
- when produce such as milk can be sold at a remunerative price.

Other Straw Feeding Systems

A wide range of additional straw feeding methods exists, ranging from selective consumption, chopping and soaking of straw, breeding or management for better or more straw, to the use of adjusted cropping patterns and types of animals. The straw is chopped and/or soaked to increase the intake and to ensure that the animals eat the entire mixture rather than select only the better parts. Breeding and management for better or more straw relate to the selection of crop types and/or management methods that favour animal production. For example, plant breeders have long overlooked that farmers in conditions of unreliable rains prefer cereal varieties that will also yield good straw in the case of harvest failure.

There are also varieties of sugar cane that produce good cane as well as fodder, a typical strategy for mixed farming conditions. Moreover, by densely planting the crops it is possible to get thinner stems, i.e. stems with higher digestibility. In a dense crop, a part of the crop can be removed before it matures, providing good feed and more space for the rest of the crop. Stripping of leaves is done to get green but maturing leaves of maize, cane and other coarse crops to be used as feed when their feeding quality is still good.

Clearly, this requires a type of animal and production level that can use such feed; the concept of the "communal ideotype" illustrated with crop and animal choice. Whatever is done with straw, it is almost always a loss of nutrients and soil organic matter when straw is burned, even though that may be the easiest form of disposal. Several methods of feeding and their usefulness for the farming system are indicated in Table 3. These depend on the mode of farming, the access to other feeds and the type and level of desired production.

EXPAGR has an abundance of forest, bush and waste land grazing, and feed resources that are a better source of nutrients than straw. Straw in these conditions is only useful to help the animals through a period of feed scarcity and selective consumption can be used. The use of straw for feed is more common in LEIA and NCA, systems that are both characterised by an adjustment of production objectives to resources. Particularly in LEIA, the shortage of feed, combined with the availability of labour and the adjustment of animal output to poor-quality feed resources, makes it relevant to chop or soak the straw in order to avoid wastage, or to make sure that a maximum number of animals is maintained.

Straw use in NCA is determined by the need to recycle better or preserve excreta, to maintain soil structure and to avoid straw burning. Biomass from crop residues can also be needed for soil protection by mulching for direct improvement of soil. Often a combination of uses exists; even competition among straw uses can occur—as animal feed, for bedding to collect urine, for the soil as mulching, for the paper industry, for roofing or for fuel.

Use of Trees for Fodder

In LEIA and HEIA there tends to be insufficient land to grow fodder as a special crop. One way to cope with a shortage of fodder is to use crop residues such as straw, as already discussed. Another approach is the use of leaves from trees that are specially grown on the boundaries of crop fields. Use of trees for fodder can also be beneficial in EXPAGR where trees grow naturally in the bush or on the ranch. Feeding of tree leaves in such systems is a common strategy to overcome periods of feed shortage.

Traditional feeding systems make maximum use of tree leaves, pods, seeds, etc. Farmers, but particularly the women, tend to be well aware of the habits and preferences of each animal. They know feed materials which are claimed to be beneficial for improving the quantity and quality of milk, and for improving the health of their animals, etc. This is a typical case where farmers use indigenous knowledge to classify feed as very good, average or bad. They do so on the basis of the palatability and visible effects of the feeds on quality and quantity of milk, unlike researchers who look mainly at chemical analysis.

Table 3. Usefulness of straw feeding methods per mode of farming in mixed crop-livestock systems

Mode of agriculture	*Relevant feeding system*
Expansion agriculture	- emergency feeding - selective consumption - catalytic supplementation
Low external input agriculture	- emergency feeding - chopping and/or soaking to avoid wastage - stripping/thinning - variability
New conservation agriculture	- straw treatment with urea - selective consumption - adjusted cropping - variability
High external input agriculture	- substitutional supplementation - straw as source of fibre in high

The research and extension community can gain much if they learn to speak the language of the farming communities.

Animal Breeding and Genetic Resources

Much concern exists worldwide about the need to maintain genetic diversity, on the one hand, and to increase the production of animals on the other. Whether goats, rabbits, chickens or cattle, many believe that the use of imported genes will increase the production of animals. Often it is even assumed that the import of genes is a miracle solution whereas, in fact, animals will only produce more if they are well taken care of, if they are healthy and well fed. Imported genes are not always improved genes even though there are conditions that call for import, often at the expense of local genetic diversity. Many traditional local breeds have already disappeared, together with their sometimes very rare characteristics such as horns, hair patterns and disease resistance, and influence on local folklore.

Fortunately it is possible to maintain the advantages of local breeds and also to reap the benefits from imported genes. Crossbreeding is the breeding of an exotic breed with a local one, aiming to maintain the best of both breeds. Where necessary, farmers and policy-makers can maintain local breeds. An example is the N'dama breed in the humid areas of western Africa that has particular tolerance to sleeping sickness. In other locations, particularly in cities with a good supply of feed, veterinarians, etc., it may be necessary to choose pure exotic breeds, whether chickens, pigs or dairy animals. Local breeds are a potential asset to countries for the adaptive traits that they have acquired over time, such as:

— tolerance or resistance to various diseases, including serious entero- and ecto-parasites;

— tolerance to fluctuations in feed resources and water supply;

— tolerance to extreme temperatures, humidity and other climatic factors.

Farmers in areas where foreign "blood" was imported through development agencies are now starting to realize that their conditions are not always suitable for the "purebreds". Breeding for production levels that suit the local resource base is related to the issue of the communal ideotype and is often done by farmers when they "go back" from exotic to crossbred or even local animals. Many livestock experts claim that environmental constraints should be removed before attention is paid to breeding because "only under favourable environmental conditions can animals express their genetic potential".

However, genetic potential is not a single magic upper limit of the performance of animals. In the first place, production should not be measured only in the number of eggs or litres of milk. Particularly in mixed systems the production should be seen as a combination of dung, draught, meat, social satisfaction, etc. Secondly, the performance of genotypes may differ between environments and, thirdly, "genetic potential" also implies an animal's capacity to cope with harsh or unfavourable conditions. Therefore the genetic make-up of stock has to be judged in the environment in which they have to produce and genetic progress can be realised even under so-called suboptimal production conditions.

Rearing of Animals

It is one thing to obtain and use exogenous breeds; it requires management to maintain healthy stock. The example of calf raising is taken here, but similar stories can be told of improvements through management in rearing goats, rabbits, chickens, etc. Many are based on hygiene, good use of critical inputs and housing, but most depend on the keen eye of the farmer. In mixed systems in particular a farmer has many things to keep in mind, and rearing of young stock can be different to that in specialised systems. Calf rearing may therefore not receive much attention and mortality can vary between 20 and 40 percent, or even higher in imported cattle.

In farms with few animals, there tends to be competition between milk for sale and for consumption by humans and milk for the calf. In dairy-oriented systems the male calf has little value if the meat of young animals does not have an extra price, such as for white or pink veal. In cropping-oriented systems, on the other hand, the male calf as a future ox for draught has more value than female calves for replacement. Climatic conditions, feeding and genotypes are important in the development of a calf over time into an adult animal. A hot climate gives lower weights at birth, a consequent lower growth rate, and lower liveweight at first service and calving.

Poor feeding results in weak calves at and after weaning with high mortality rates, which in turn leads to little development of the mammary system, to a high age at first calving and to a low production of milk from the adult animal. Crossbreeding in particular results in a high variability of growth and corresponding liveweights and there is a wide range of weights at all ages at smallholder level for the combination of the above factors.

Conditions can be improved by providing shade and natural and/or artificial cooling. Early feeding of hay and concentrate can correct poor milk feeding practices. The use of the adult weight of a particular crossbreed can help to determine the correct birth weight and growth rate one can expect. For example, a 300 kg crossbred cow is expected to produce an offspring weighing between 6 and 8 percent, i.e. 24-32 kg, with a daily growth of 300 g/d. From a 500 kg crossbred one expects a 30-40 kg calf with a growth rate of 500 g/d.

A well-reared calf is the productive cow of the future and various approaches have been tried to improve calf rearing. In dairy regions of the United States and western

Europe there are calf-rearing clubs where farm children rear a calf in the best way they can while they register the costs of their exercise. An animal husbandry inspector regularly judges the technical and economic performance of the calves at the farm or a central place in the region while giving comments on the youngsters' efforts, and allocates prizes. Similarly, cooperatives in India, Sri Lanka and eastern Africa have held calf rallies based on the principle of judging at a central place.

Most dairy textbooks advocate artificial rearing to reduce the amount of milk consumed by the calf and leave more for family consumption and/or for sale. This approach is borrowed from conditions where artificial milk is cheaper than milk "direct from the cow". In many smallholder conditions the situation is different. Artificial milk is not easy to obtain, and restricted suckling is a good option. Restricted suckling is a practice in which a calf is allowed to suckle after the mother is milked; it reduces or prevents mastitis, a problem common in herds with artificially reared calves. It is also very convenient for the cow in the case of poor milkers and in cases where the farmer has no time for milking, or when he/she wants to attend meetings away from the farm.

Bucket feeding is more troublesome, especially with calves that carry zebu blood, because they cannot easily drink from the bucket. In addition, feeding with artificial milk powder can be troublesome due to low-quality powder, difficulty in getting hot water and cleaning practices that are below standard.

Technoloites for Meat and Milk Processing

Animal products such as milk and meat are difficult to keep over long periods. They can spoil and become unsuitable for human consumption in a matter of hours, particularly in hot and unhygienic conditions. A range of preservation techniques can be found to process this category of products, including heating, smoking, salting, fermenting, drying or concentration to reduce the moisture content of the raw materials. For example, dried, smoked or salted meats are prepared to preserve the meat and to change the flavour and texture to increase variety in the diet.

Conditions under which animals are slaughtered are often unhygienic and allow bacteria to be transmitted to meat by flies, animals and birds. Harmful micro-organisms may grow on meat and cause food poisoning when eaten. These, together with infectious organisms such as parasites that grow in the meat, make careful selection of raw materials, proper handling, hygiene and preparation of meat products essential. In the case of milk, many traditional techniques can be found for producing indigenous products. Fermented milk products such as yoghurt and soured milk contain bacteria that aid digestion and that help prevent illness caused by other bacteria. In addition, fermentation removes milk sugar from milk and facilitates digestion of the product.

Examples of rural processing of meat products include dried, salted and smoked meats. They are found widely in Africa where they are important traditional foods. Possibly the best known example is biltong, which is used as a snack food in southern Africa, but similar products are made in other parts of the continent. Biltong is made from strips of dried, salted meat, which are dark brown with a salty taste and a flexible, rubbery texture. Quanta is a similar dried and spiced beef product used as snack food in north Africa. Spices are mixed with fermented honey and about 120 grams of spices are added to each kilogram of meat and mixed in manually. The evenly spiced strips are hung on a string suspended in a well ventilated, dust-free area and left to hang for five to seven days until they break with light hand pressure or are crunchy upon chewing.

Examples of dairy products are soured milk, cheese and butter. Traditional sour milk is a thick clotted product similar to yoghurt but with a stronger flavour and a more acidic taste. It has a similar shelf life of three to eight days and is used as a drink or as an accompaniment to meals in some countries. Preservation is due to the production of lactic acid by naturally occurring lactic acid bacteria in the untreated milk. Various types of fresh cheeses can be made with simple equipment; a pan or container and some lemon can be good enough to prepare simple but tasty cheese. Traditional butter and ghee are made by stirring sour milk until the fat coagulates and separates into the solid butter and liquid buttermilk. The butter is collected by hand and washed with clean water two or three times before packing. This butter can be heated to evaporate the remaining water. The result is ghee, a highly valued cooking and frying ingredient, particularly in Asia.

6

Characterisation of Mixed Farming

Mixed farming is common worldwide, in spite of a tendency in agribusiness, research and teaching towards specialised forms of farming. Obviously, mixing has both advantages and disadvantages. For example, farmers in mixed systems have to divide their attention and resources over several activities, thus leading to reduced economies of scale. Advantages include the possibility of reducing risk, spreading labour and re-utilising resources. The importance of these advantages and disadvantages differs according to the sociocultural preferences of the farmers and to the biophysical conditions as determined by rainfall, radiation, soil type and disease pressure.

Mixed farming exists in many forms depending on external and internal factors. External factors are weather patterns, market prices, political stability, technological developments, etc. Internal factors relate to local soil characteristics, composition of the family and farmers' ingenuity. Farmers can decide to opt for mixed enterprises when they want to save resources by interchanging them on the farm—because these permit wider crop rotations and thus reduce dependence on chemicals, because they consider mixed systems closer to nature, or because they allow diversification for better risk management.

There is wide variation in mixed systems. Even pastoralists practise a form of mixed farming since their livelihood depends on the management of different feed resources and animal species. At a higher level, a region can consist of individual specialised farms and service systems that together act as a mixed system. Other forms of mixed farming include cultivation of different crops on the same field, such as millet and cowpea or millet and sorghum, or several varieties of the same crop with different life cycles, which uses space more efficiently and spreads risks more uniformly.

The study of a wide variety of mixed systems at different levels is beneficial to understanding the logic of mixed systems in general. Disciplines such as ecology, economics and complex system theory have tools and concepts that can help us to understand better the mixed blessings of mixed systems. One essential point here is that the principle of mixing occurs everywhere, also in society—domestic waste such as glass, bottles or paper is also recycled. Another point is that in mixing the different functions of plants and animals can be observed: a cereal crop produces grain and straw, a legume provides grain, organic matter, fodder and nitrogen. A third point is that it tends to be more important to look for high yield of the combination of the components rather than for the yield of one component.

Mixed farms are systems that consist of different parts, which together should act as a whole. They thus need to be studied in their entirety and not as separate parts in order to understand the system and the factors that drive farmers and influence their decisions. That principle is here referred to as the "command ideotype". It may be the most important principle to achieve increased production in mixed systems, together with the awareness that crops and animals have multiple functions.

Types of Mixed Farming

Mixed farming systems can be classified in many ways—based on land size, type of crops and animals, geographical distribution, market orientation, etc. Three major categories, in four different modes of farming, are distinguished here. The categories are:

- On-farm versus between-farm mixing
- Mixing within crops and/or animal systems
- Diversified versus integrated systems

The modes of farming refer to different degrees of availability of land, labour and inputs, ranging from plenty of land to a shortage of land. The modes are characterised by Schiere and De Wit as expansion agriculture (EXPAGR, plenty of land), LEIA, HEIA and new conservation agriculture.

On-farm mixing refers to mixing on the same farm, and *between-farm mixing* refers to exchanging resources between different farms. On-farm mixing occurs particularly in LEIA where individual farmers will be keen to recycle the resources they have on their own farm. Between-farm mixing occurs increasingly in HEIA systems—in countries such as the Netherlands it is used to mitigate the waste disposal problems of specialised farming. Crop farmers use dung from animal farms, a process that involves transport and negotiation between farmers and even politicians. Between-farm mixing also occurs at the regional level—in the store cattle systems of the United Kingdom and the United States, animals are raised in one area to be fattened in another area where plenty of grain is available. In tropical countries also, manure may be transported from livestock farms to farmers and vegetable cropping areas where manure is in short supply.

Pastoralists from such systems in West Africa and on the Indian subcontinent also exchange cattle and crop products with crop farmers. Cultivators receive manure, labour and, less important, milk in return for cash, grain and water rights traded to pastoralists. Entrustment of livestock from crop farmers to pastoralists follows more or less the same rules. In return for taking care of the herd, herders receive either cash, or cropland, or labour for the cropland or a share of the milk and the offspring.

Mixing between nearby farms is considered here as providing the same advantages as on-farm mixing, but it should be underlined that there are important differences in terms of social organisation and transaction costs. For example, in West Africa the exchange between farms leads to tension and accidents as crop farmers start to use land that used to be pastoral only. The amount of grazing land is decreasing and dependence on the grazing of crop residues is increasing. When herders bring in the animals before the field is properly harvested serious incidents and conflicts can arise.

Mixing within Crop and/or Animal Systems

Mixing within crop and/or within animal systems refers to conditions where multiple cropping is practised, often over time, or where different types of animals are kept together, mostly on-farm. Both these systems occur frequently though they are not always apparent. Within-crop mixing takes place where crop rotations are practised over and within years. For example, a farmer has a grain-legume rotation to provide the grain with nitrogen or a potato-beet-grain rotation to avoid disease in the potatoes. Plants can also be intercropped to take maximum advantage of light and moisture, to suppress weeds or prevent leaching of nutrients through the use of catch crops. Examples of mixing between animals are found in chicken-fish pond systems where chicken dung fertilises the fish pond; in beef-pork systems where pigs eat the undigested grains from the beef cattle dung; or in mixed grazing such as cow-sheep mixes to maximise biomass utilisation or to suppress disease occurrence.

The distinction between diversified and integrated systems is perhaps the most relevant one for this report. Diversified systems consist of components such as crops and livestock that co-exist independently from each other. In this case the mixing of crops and livestock primarily serves to minimise risk and not to recycle resources. Integration is done to recycle resources efficiently. It occurs in mixed ecological farms of temperate countries, but also in mixed, relatively low input farms of southern and southwestern Australia with grain-legume-sheep mixtures.

Integration occurs most often, however, in LEIA farming systems that exist in many tropical countries where products or by-products of one component serve as a resource for the other—dung goes to the crops and straw to the animals. In this case the integration serves to make maximum use of the resources. Unfortunately, these systems tend to become more vulnerable to disturbance because mixing of resource flows makes the system internally more complex and interdependent.

In Asia, the integration of livestock, fish and crops has proved to be a sustainable system through centuries of experience. In China, for example, the integration of fishpond production with ducks, geese, chickens, sheep, cattle or pigs increased fish production by 2 to 3.9 times, while there were added ecological and economic benefits of fish utilising animal wastes. Environmentally sound integration is ensured where livestock droppings and feed waste can be poured directly into the pond to constitute feed for fish and zooplankton. Livestock manure can be used to fertilise grass or other plant growth that can also constitute feed for fish. Vegetables can be irrigated from the fishponds, and their residues and by-products can be used for feeding livestock.

Grazing of livestock under plantation trees such as rubber, oil palm or coconut is a form of crop-livestock integration that is often found in Southeast Asia. Experiments in Malaysia with cattle and goats under oil palm showed better oil palm bunch harvest and comparable results were found where goats fed under rubber trees. In rubber and oil palm plantations in Malaysia, the integration of livestock to utilise the vegetative ground cover under the tree canopy increased overall production and saved up to 40 percent of the cost of weed control. Similarly, sheep helped to control weeds in sugar cane fields in Colombia. This suppressed the costs of herbicides, reduced the cost of weed control by half and provided additional income from meat production. This also occurs where cows graze under coconuts.

The best known type of integrated mixed farming is probably the case of mixed crop-livestock systems. Cropping in this case provides animals with fodder from grass and nitrogen-binding legumes, leys, weeds and crop residues. Animals graze under trees or on stubble, they provide draught and manure for crops, while they also serve as a savings account (Figure 1).

Modes of Mixed Crop-livestock Systems

Even in integrated systems the exchange of resources such as dung, draught and crop residues takes place in degrees that differ among the so-called modes of farming, based on the availability of land, labour and capital respectively (Table 1):

— Expansion agriculture (EXPAGR)

— Low external input agriculture (LEIA)

— High external input agriculture (HEIA)

— New conservation agriculture (NCA)

The EXPAGR mode occurs where land is abundant, i.e. where shortage of land or local fertility are overcome by migration or by expansion into other regions where bush and forest fallow still occur. Typical examples of mixed farming in this mode are found in West Africa and in old Asian and European grazing systems. Animals were sent out to graze and would come home to "pull the plough or fertilise the crop fields". The crop fields themselves could move elsewhere if local soil fertility declined. However, this mode is becoming more rare as land resources are exhausted throughout the world.

Mixed farming in LEIA occurs where the shortage of land can no longer be overcome by migration or use of substantial areas elsewhere for grazing. Lack of access to external inputs such as fuel, chemical fertilisers or pesticides implies that only increased use of labour and skills offers a way out. This also implies the introduction of modified practices, and the need to adjust demand according to resource availability. Dung is carried around on the farm by using more labour because a lack of soil fertility cannot be compensated by shifting to more land or by employing more livestock to "produce" more dung.

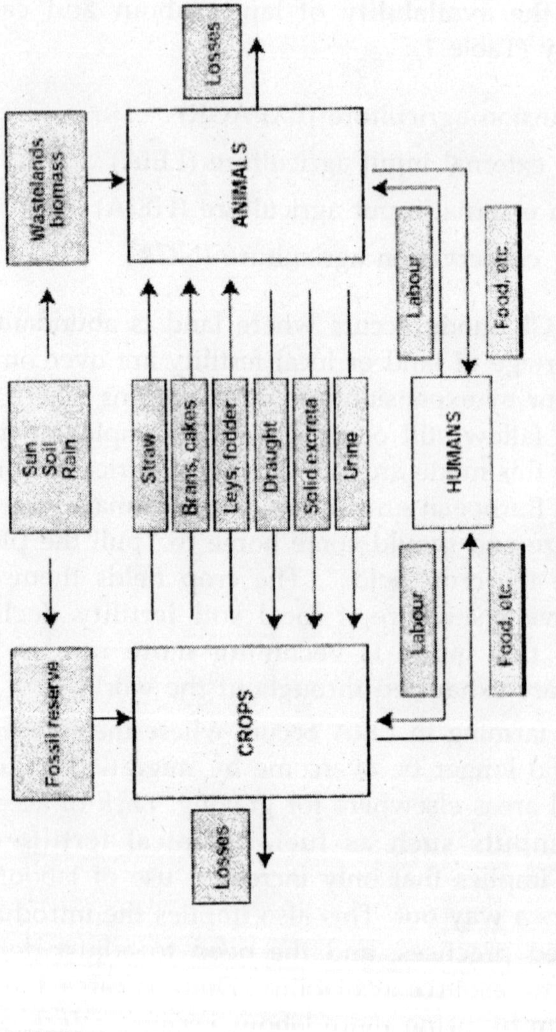

Figure 1. Different modes of mixed farming

In LEIA systems the latter is considered a resource but a waste product in HEIA systems. If not managed properly and if demand for food and other crops is not adjusted to the carrying capacity of the soil, this can result in mining of soils and/or collapse of the systems. Van der Pol calculated that the cotton-cereal systems in southern Mali earned 40 percent of their income by mining the soil. However, this cannot go on indefinitely and sooner or later the system will collapse if there are no changes. Some researchers think that animals, when managed correctly, can serve to fill part of the gap that exists between the output and the input of nutrients in the system, together with a proper use of chemical fertiliser.

Mixed farming in the HEIA mode is not frequently found because it implies plentiful access to resources such as external feed and fertiliser that make exchange and recycling of resources at farm level not relevant. Exchange of resources between farms only exists, as seen in the section On-farm versus between-farm mixing, after the excessive use of fertiliser forces farmers to recycle the waste. In the HEIA mode the demand for output determines the use of inputs. The use of external resources can reach such high levels that the environment is affected by emissions from the crop and/or animal production systems, ultimately leading to waste disposal problems, thus forcing HEIA into NCA.

NCA is a mode of farming where production goals are matched as closely as possible to the resource base. This approach represents a mix between HEIA and LEIA, i.e. it aims to replace the removed nutrients but it also aims to achieve keen farming and adjusted cropping and consumption patterns to suit local conditions. The use of leys is important to regenerate soils, to add nitrogen, to mobilise phosphate and to suppress weeds.

Table 1.

Characterization of different modes of mixed crop-livestock farming

Mode of farming	EXPAGR	LEIA	HEIA	NCA
Relative access to production factors:				
Land	+	-	-	-
Labour	-	+	-	+/-
Capital	-	-	+	+/-
Characteristics of farming:				
Source of animal feed	Outfield	Infield roadsides	Infield Import	Infield
Role of animals as savings account	High	Medium	Low	Low
Importance of excreta				
- Dung	Positive	Positive	Negative	Positive
- Urine	Neglected	Positive	Negative	Positive
Source of energy for labour	Humans/animals	Humans/animals	Fossil fuel	Fossil fuel/animals
Form of mixing	Diversity	Integration	Specialization	Integration
Can be between and on-farm	On-farm	May be between farms	Mainly on-farm	
Crop residue feeding	Irrelevant	Very relevant	Irrelevant	Relevant

Contd....

Contd....

Role of leys				
- For weed control	NA	Low/NA	NA	Important
- For nutrient dynamics	NA	Low/NA	NA	Important
- For erosion control	NA	Low/NA	Low/NA	Important
Ratio outfield/infield				
- Local level	High	Low	Low	Low
- International level	Low/NA	Low/NA	High	Low/NA
Output of milk or meat per animal	Low	Low	High	Medium
Attention to conservation of the resource base	Low	Medium	Low	High

Traditional Technologies

Some technologies are useful to an individual, others to a group of farmers, some are inherently more sustainable than others, and some depend on use of more inputs while others can help to reduce dependency on external inputs. Some technologies can be called exploitative; others are regenerative, as explained in the following section. The application of technology is not necessarily progress, but may be a response to a shortage situation. For example, in EXPAGR with ample feed available there is no incentive to plant fodder unless the existing fodder base runs out. In that case the farmers start to feed straws for lack of grass (LEIA), or they start to grow fodder based on fertiliser or legume-nitrogen (NCA).

Farmers will not be convinced to vaccinate or house the animals if there is no danger of disease or predators; attention to veterinary aspects is a necessity, not a luxury. Then again, one technology may serve several goals at once and it may also carry disadvantages. Housing helps to protect the animals, to facilitate daily chores or to conserve feed. It also costs money and resources, it constrains the freedom of the animals and it may even negatively affect the health of the animals. Similarly, there is hardly any technology that comes alone. The use of a milking machine requires the availability of skilled technicians and training of a farmer; the introduction of purebred pigs requires more elaborate housing, arrangements for supply of genetic material and availability of veterinary care. In some cases even the local food habits may have to change—for example, when slow growing pigs with much fat are replaced by fast growing pigs that have relatively lean meat and that use certain kinds of feed more efficiently.

Technology in general can be defined .as a method to overcome a problem. A farmer getting tired of milking the cow will be happy if there is a labourer or a machine to do it instead. Farmers who get tired of hand weeding the crops will be happy to use a hoe or chemicals to make the job easier. Any intervention can be considered a technology—whether it is technical, management or policy related—but there is one condition. The cost of the technology should be paid by a combination of the extra return and the saved effort, a "calculation" that tends to be done differently among farmers themselves, scientists and/or policy-makers.

Types of Technology

Several different technologies exist in relation to mixed farming and the following distinctions are made:

— versus management-based technology

— Accelerating versus defusing technology

— Exogenous versus indigenous technologies

— Technologies for national and local problems

— Technologies for individual farmers and for society

— Exploitative and regenerative technologies.

As in any other classification there is always another way to distinguish between types of technology. The purpose of this classification is to show major differences. The interventions discussed later will, in practice, always be a mixture of those mentioned above.

Input- and Management-based Technologies

Input- and mechanisation-oriented technologies in -c rease the output of a particular animal or farm by using more inputs (feed, fertiliser, pesticides), by using a machine to save labour, or by digging wells if there is not enough water. Management-based technologies are focused on trying to understand the farm as a combination of soil, plants or animals; the region as a combination of farms, people, mountains, underground water reservoirs, etc. Proper management can help to avoid losses where possible. For example:

— The cow when it stands up in the morning takes some time to stretch and to urinate/defecate. Knowing this, the farmer waits for fifteen minutes before letting the cow out so that the dung and urine are collected in the animal shed.

— When building the sheds, the housing can be oriented either to let the sun shine in unnecessarily and at the wrong moment, or the sun enters at times when it suits the management.

— Much effort can be spent on collection and preservation of animal manure but appropriate management is required to apply it at the appropriate moment to serve its purpose.

— Veterinarians can be called in to cure disease but proper management (hygiene, timely feeding) and a keen eye (noticing the onset of a disorder) can help to prevent the need for a cure.

Accelerating and Defusing Technologies

The distinction between accelerating and defusing technologies is difficult, but it is related to the earlier distinction between input and management oriented technologies, and to the exploitative-regenerative technologies. An example of an accelerating technology is when a cow is given a medicine that "forces" the animal to become pregnant even when the cow's body is too weak to conceive. By forcing it to conceive again the cow will collapse and even more inputs are required to get the cow in shape again. Liming of the soil or cultivation of *Brachiaria* or cassava on poor soils can also be considered as accelerating technologies.

Brachiaria is a fodder grass and cassava is a tuber crop for human consumption, but they have in common that they both grow well (initially) on poor soils. However, they exhaust an already poor soil even more. The same process takes place when goats or cattle are left to graze already eroded hillsides. Liming, also called marling, was a well-known form of fertilisation in Europe for many centuries. It was done, among other reasons, to release phosphorus from the soil-nutrient complex by adding calcium. Ultimately it led to soils being depleted, as evident from this example of old British farmers' wisdom of several centuries ago:

> "It brings the grounde to be starke nought, wherby the common people have a speache, that ground enriched with chalk makes a riche father and a beggerly sonne."

or in more common language:

> "Lime and lime without manure, makes both land and farmer poor."

Defusing technology as opposed to accelerating technology tends to consider the problem to a greater depth, to look for its root causes and see how the basic processes can be stalled or reversed. Defusing technologies are relevant in NCA and they tend to be approaches that fallow land, that rotate the crops rather than treating them with chemicals to enhance fertility or to combat weeds / disease, or that reduce erosion rather than living with the consequences of erosion. By accepting natural limitations the defusing technologies would use tolerant breeds such as N'dama or Baoulé in areas with sleeping sickness, where otherwise heavy control measures would be required. When a cow is weak a diffusing management practice would be to stop milking her a few weeks sooner to help her rebuild the reserves necessary for the next lactation. Ultimately, the diffusing management approach implies attention to stress signals from the environment. It aims to reduce or adapt consumption and to restore the resource base rather than to aim beyond carrying capacity.

The distinction between indigenous and exogenous technology is again not strict but it serves to make a difference between solutions generated and collected over many generations by farmers themselves and solutions that come from outside. By stressing the existence of indigenous technology one unlocks a vast potential of local knowledge and creativity that helps to modify external technologies for local use. Farmers can often come up with cheap solutions that would not have been generated from outside. Technical solutions from outside can certainly be useful—a vaccine against disease, a new way of conserving feed, etc. Others, however, do not fit the local conditions and it may be better to use a local and cheaper solution.

Typical examples of indigenous knowledge are ethnoveterinary medicine, where women know the local herbs to cure or prevent disease in their goats, chickens or other animals, and knowledge about the best timing for ploughing or planting. Use of indigenous technologies can also be relevant for coping with variations in soils, herding rights, local business people, etc. Indigenous knowledge is no "cure for all", e.g. it has no or almost no ways to cope with infectious diseases and occurrence of new disorders that come with development. Still, the use of local knowledge for local problems has been shown to improve the uptake of development programmes substantially.

Technologies for National and/or Local Problems

Some problems in animal production are felt more at national than at local farmers' level. Rinderpest, foot-and-mouth disease (FMD), classical swine fever, etc. cause damage for the farmers, but they mostly threaten the export licences of countries. In other words, many farmers will not feel motivated to do things that are not in their own direct interest. Large-scale programmes will have to take that into account. A more dramatic example is when governments want food to be cheap and plentiful for urban populations. This may require the production of cereal varieties that are not liked by the farmers themselves. It may even force governments to work against local traditions that, for example, consider milk as a gift from the gods that is not to be traded for profit. Improper handling of such tensions will not lead to effective development.

Technologies for Individual Farmers

Cooperatives are useful forms of farmers' organisations for development, for example to supply inputs, to work on credits and savings, or to work on a watershed programme. However, not all technologies are useful for collective action. For example, a vaccination campaign almost certainly needs the organisation of groups, whereas the treatment of a broken leg or a difficult birth in an animal is based on individual interests. Farmers' study groups are collective action but farmers' investment decisions and cropping/livestock keeping strategies are the decisions of the farmers and/or their extended families themselves.

Exploitative and Regenerative Technologies

This distinction resembles that under the section *Accelerating and defusing technologies,* but it treats the time horizon of the farmers and of farming. Planting of *Brachiaria* leads to rich yields but also to more exhausted soils. The planting of legumes or investments in soil regeneration give lower short-term yield but ensure a livelihood for the future generation. In particular, the mixed farming of the NCA mode puts heavy emphasis on regeneration by involving technologies such as the use of leys, recycling of nutrients, mutual adjustment of cropping/livestock components and erosion control through fodder crops.

Concept of Communal Ideotype

One particular issue in the suitability of technologies for mixed systems is in the concept of the "communal" ideotype. This term was coined by Donald for wheat breeding and it emphasizes that the yield of an individual plant needs to be made subject to the yield of the entire plot. More practically, no farmer is interested in high individual plant yield if it does not lead to the increase of total farm yields. The principle of the communal ideotype and choice of technologies for the design and choice of technologies in mixed crop-livestock farms can be illustrated in many ways:

— In fodder crops it has been shown that napier grass and leucaena can be grown alone or in combination, particularly for NCA conditions. Growing of only one crop gives higher yields per crop but a lower yield from the combination (Table 2).

— A study for LEIA conditions in India with different crop-animal combinations showed that the use of cows with individual productions of 10 litres/animal/day led to higher total farm outputs than those producing 16 litres/animal/day (Table 3). The high- producing dairy cows cannot be fed with fibrous crop residues such as straw, i.e. the goal of high milk yield per cow would reduce total farm income, thus losing an opportunity to use these resources.

Table 2. Yields of napier (Pennisetum purpureum) and leucaena (Leucaena leucocephala) in pure and mixed stands

	Yield (tonnes/ha/yr)		
	Napier	Leucaena	Total
Napier alone	12	-	12
Leucaena alone	-	8	8
Napier + leucaena	10	6	16

Table 3. Optimum crop combinations, herd size and production at different individual cow productions with or without treatment of stover, when the farmer also has access to a small fixed area of good quality fodder

Individual production	Total production (litres/day/ system)	Herd size (cows/ farm)	Cotton(ha)	Total income from milk and crop (litres/day/ cow) (Rs/day/farm)
2.0	5.1	2.5	0	22.2
6.0	9.5	1.6	0	35.4
10.0	10.6	1.1	0.4	39.1
16.0	6.6	0.4	1.0	27.6

— A study of crop-livestock combinations in Kenya shows an added issue, i.e. the optimum combination of animals and crops. The large farmer benefits more from grade cows and the small farmer is better off with crossbreeds (Table 4).

Table 4. Optimal farm crop areas as calculated with linear programming for an area with mixed crop-livestock systems in Kenya

	Large farm holding	*Small farm holding*
Gross margin (Kenyan shilling)	7 952	6 560
Land used (ha)	7.98	3.9
Grade cows (cow/farm)	7	-
Crossbred cows (cow/farm)	-	5
Coffee (ha)	0.06	-
Maize (ha)	0.28	0.52
Beans (ha)	1.03	1.34
Potato (ha)	0.03	0.04
Banana (ha)		0.15

Field observations and common sense support the point that farmers adjust parts of their farms to achieve larger overall output. In many tropical and temperate contexts they use crossbred cows or medium-producing animals rather than purebreds. In drought-prone areas farmers choose grain crops that also yield good straw/fodder, even at the expense of some grain yield.

7

Improving Water Resources

Advancing Restricted Rainfall Infiltration

Infiltration depends on there being sufficient porosity in the surface soil for rainfall to infiltrate, and in the subsoil and parent material for rainwater to percolate. When the porosity of the surface soil is too low to accept rainfall, or subsoil porosity is too low to allow rainwater percolation (i.e. permeability is too slow), then infiltration will be restricted and rainwater will be lost as runoff. The porosity of surface soil may have been reduced by clogging of pores with particles detached from soil aggregates under the impact of raindrops, or by the deposition of detached particles on the soil surface as impermeable crusts or seals.

The porosity of subsurface soil may be naturally low, or may have been reduced by compaction and tillage practices that have disrupted or destroyed pore spaces causing a zone of low permeability at the base of the tilled layer. The degree to which soil porosity is reduced by tillage is frequently sufficient to limit root penetration, but is less often so severe that permeability to water is significantly diminished. The over-riding approach should be to instil in society, and in farmers, extensionists researchers in particular, the will to create and sustain soil conditions that encourage the infiltration of rainfall where it falls, and to counteract the causes of runoff. This implies that the porosity of the soil must be at least maintained, or increased.

Infiltration Capacity of the Soil Surface

Porosity of the soil surface is best maintained by first protecting it from the disruptive action of raindrops through a protective cover, usually of residues from the previous crop, a cover crop or mulch, and by ensuring the soil is not disturbed by tillage. The effects of conservation agriculture on higher infiltration and reduced runoff and flooding have been well documented in Brazil in particular. If the whole concept cannot be applied immediately, improvements in soil moisture status of the soil can still be achieved, though probably not to the same extent, by other measures aimed at prolonging the useful life of rainwater. These include the use of surface residue covers alone, fallow periods under cover crops or natural vegetation, protection or temporary closure of grazing lands and forests from overgrazing, and operations on the contour, complemented by physical measures to detain rainwater.

The regular use of shallow tillage with disc or tined implements to break-up surface crusts to increase surface porosity and enhance rainfall infiltration is not recommended. The increase in surface porosity is only temporary and on crusting-susceptible soils tillage will need to be repeated after every rainstorm. Tillage leads to the disruption of pore spaces in the soil, and the use of discs, in particular, often causes compaction, which may impede root growth and rainwater percolation. Tillage also accelerates the loss of soil organic matter leading to a progressive deterioration of soil architecture and a reduction in the number and stability of pores that allow growth of roots and movement of rainwater. Regular tillage therefore is not recommended as a solution to restricted infiltration caused by low porosity of the soil surface.

Residue Cover

A residue cover absorbs most of the energy of the raindrops that fall on it and by the time this rainwater reaches the soil below, its ability to disintegrate soil aggregates and detach fine particles is greatly reduced. Consequently, there is little or no clogging of surface soil pores by detached particles, and little deposition of soil particles that would form a crust on the surface. The benefits of a residue cover are most apparent on soils initially in reasonable physical condition, but even under these conditions runoff can sometimes occur despite a good soil cover. For example, runoff will occur when rainfall intensity is greater than the soil's infiltration rate, or when the soil's pore spaces are already filled with water because the soil is shallow, its water holding capacity is low, or its subsoil is only slowly permeable.

When a residue cover is applied to a soil with a very degraded surface of low porosity, the beneficial effect of the cover on infiltration may be initially limited. In such situations, it is advisable to accelerate the recuperation of surface porosity before applying residue covers by tilling the soil once to break-up the crust and any subsurface pans, followed by a fallow period under a cover crop to enhance the formation and stabilisation of soil porosity.

The choice of a cover material depends on what is locally available. Residue covers may consist of:

- Crop residues left in the field after harvesting the previous crop.

- — Cover crops sown the previous season and left on the soil surface after slashing or applying a herbicide.
- — Leaves and branches lopped from trees growing within the cropping area.
- — Mulches of grasses, shrubs, weeds, litter, husks and other organic waste materials.

Improvements in crop yield and quality of soil occurred and a high proportion of farmers adopted these measures. Although erosion control was cited as the farmers' main reason for not burning crop residues any more, an important pointer to the benefits due to improved soil moisture conditions was evident in 1997. In that year there was a serious drought during the rainy season associated with the El Niño weather phenomenon. But according to the farmers, they were able to harvest almost as much maize as in a more normal year because of conservation of moisture in the soil as a consequence of the better soil status, while neighbours who had not adopted the system lost their crops to drought.

Nor did they lose their crops the following year during hurricane Mitch and the associated torrential rainfall, which caused disastrous flooding. The farmers noticed that the same mulch prevented the seeds from being washed away by rainstorms and facilitated rainwater infiltration so that they did not have problems of decaying plants during the heavy rains. A cross-check on this beneficial effect under the same extreme weather conditions comes from Honduras, where hurricane Mitch caused much erosive devastation on many hillsides, but less on those hillsides where soils were well protected by crop residues.

In a limited area of western Honduras the Quesungual traditional agroforestry system has been used by small farmers to produce maize, sorghum and beans. As rising population pressure makes the traditional slash-and-burn system increasingly unsustainable, there is increasing interest among farmers in the Quesungual system. It combines pruning of naturally regenerating indigenous trees and shrubs with normal agroforestry methods for growing high-value timber and fruit-trees. Before sowing, vegetation is cut down by hand without burning and is spread across the field together with the branches and leaves from pollarding. Crop seeds are then scattered or jab-planted with a stick through the mulch layer. Weeds are cleared infrequently, by hand or using a herbicide.

Mechanisms by which surface residue covers enhance rainwater infiltration. The physical contacts between a residue cover and the soil surface obstruct the movement of the runoff, slowing it down, giving more time for infiltration and so reducing the volume of runoff. Thus two aspects of surface cover can be distinguished:

— all surface cover absorbs the energy of raindrops and so prevents the loss of pore spaces into which rainwater can infiltrate;

— contact cover slows down any runoff, giving more time for infiltration.

The degree of contact cover is important especially on steep slopes, on soils with naturally low infiltration rates, and on degraded soils with surface crusts or seals of low porosity. Furthermore, it is the contact cover that is immediately accessible to soil macro-organisms and can stimulate their activity. Thus greater numbers of biopores are likely to be formed, leading to more rapid infiltration and percolation. This is why major disturbances such as tillage or incorporation of residues, mulches or other organic matter drastically reduces these positive effects.

Pliable materials of short length, such as leaf or grass mulch, which can be easily flattened by raindrops, will develop a high degree of contact cover and will substantially slow down the speed of runoff flow, generally resulting in reduced volumes of runoff. In contrast, inflexible long materials, such as woody branches of tall bushes that are not easily flattened by raindrops, will develop a low contact cover and so have less influence on the speed of runoff flow.

Benefits of Surface Residue Covers

The advantages of mulches are the same as for crop residue, i.e. increased infiltration, decreased runoff, and greater soil water availability. They both provide additional benefits, notably less soil water losses by evaporation, less weed incidence and water losses by transpiration, softer and more workable soils, increased earthworm activity, the incorporation of additional nutrients and frequently increased yields.

In western Kenya, mulching with Tithonia has given substantial yield increases of maize, kale, tomatoes and French beans. Net profits from mulching kale ranged from US$91 to US$1 665 per ha. In the semiarid zone west and north-west of Mount Kenya, maize yields increased by a factor of 4.4 when 3 t/ha of mulch were applied. Termites were not a problem in this area, probably because of the cool climate.

Limitations of Surface Residue Covers

The main disadvantage of applying mulch is the cost or labour of collecting, transporting and applying the mulch. This is not the case with crop residues, which are produced on-site. Often, there will be no suitable mulching materials in the vicinity of the farm, or there is insufficient labour available. Transporting large quantities of mulch for large-scale cropping is seldom economic and mulches cannot be applied after emergence to closely spaced crops. When a cover crop is used as mulch, there is the cost of slashing the cover crop or applying a herbicide. Similarly, lopping trees and distributing the branches and leaves over the cropping area requires considerable labour. On steep slopes, the application of residue covers is not easy and requires much labour as well. Moreover, these materials are easily washed downhill on steep slopes.

Mulching materials and crop residues are often grazed by cattle belonging to the farmer, the community or the landowner (in the case of tenant farmers), fed to livestock, or sold as fodder. Sometimes these materials are in demand for thatching or fuel; in many semiarid areas they are rapidly consumed by termites, and in hot humid climates, they decompose rapidly. Another disadvantage of mulches is a progressive decrease in soil fertility where the mulching materials are produced, unless manures or fertilizers are applied. In parts of Uganda, the residues of cereals grown on hillsides are used to mulch bananas on the lower slopes or valley bottoms, which become enriched in nutrients at the expense of the cereal areas. Soil erosion may also degrade the source areas when the cover provided by the vegetation is removed for use as mulch.

Factors for Adopting Surface Residue Covers

The use of soil covers is more common in subhumid and humid zones because of the greater availability of vegetative materials. Nevertheless, they are particularly suited to semiarid areas when materials are available and in the absence of severe termite problems. Mulches are often applied to limited areas of high-value horticultural crops and home gardens in easily accessible fields with gentle slopes.

Cover Crops Under Uncultivated Soil

When soils are so badly degraded that they must be taken out of production, soil porosity can be restored through the action of biological processes. This can be achieved by fallowing for 1 or several years under natural vegetation, natural vegetation enriched with fast-growing leguminous trees, or planted fallows. The accumulation of large amounts of biomass on the soil surface from the fallow vegetation associated with high biological activity and strongly developed root systems promote the biological recuperation of soil porosity. Biological incorporation of residues into the surface soil results in higher soil organic matter in the upper few millimetres, which progressively extends into deeper layers overtime. The permanent cover of surface residues encourages soil faunal activity, which combined with higher soil organic matter contents leads to improved soil porosity.

A well-adapted, deep-rooting leguminous cover crop often speeds up the recuperation of soil porosity compared with a natural vegetation fallow because larger amounts of biomass are rapidly produced by the cover crop. Whereas a natural vegetation fallow may require 3-5 years, a cover crop may recuperate soil porosity in 1 year. When degraded soils are severely compacted, deep tillage with a subsoiler immediately prior to sowing the cover crop encourages establishment and development of the cover crop. If the degraded soil is severely deficient in phosphorus the application of P fertiliser will be necessary to encourage the establishment of the cover crop.

A constraint of soil recuperation by natural vegetation fallows in mechanised production systems is the problem of eliminating trees and excavating roots before returning to cropping. If a manual system is to be adopted, the problem is less serious. Herbaceous and shrubby cover crops can be eliminated much more easily by slashing, mowing or application of a systemic herbicide, and the subsequent crop may be sown directly into the residues of the cover crop.

Temporary Closure of Grazing Lands and Subsequent Protection

Low infiltration and high runoff can occur on grazing lands even on slopes less than 2 percent, as for example at Sebele, in Botswana. In this area, vegetation cover was considered to be the most important factor controlling infiltration and runoff, and catchments with a cover in excess of 70 percent generally had lower runoff compared with those with less than 70 percent cover. Although the percentage of grass cover in grazing lands has an important influence on rainfall infiltration, soil surface porosity can be more important, especially when overgrazing has degraded the soil, resulting in surface compaction and very low porosity.

On degraded grazing lands at Iiuni, Kenya, for example, even with 57 percent vegetative cover the runoff was in excess of 60 percent. The presence of algae growths on bare surfaces that were resistant to wetting encouraged runoff, whereas stone covers reduced runoff due to the creation of water-storage areas between the stones where the rainwater is detained, allowing more time for infiltration.

Forest Protection for Water Infiltration

Forest provides an excellent protective cover made up of the canopy, low-storey bushes, herbs and surface litter, which combine to protect the soil surface from loss of porosity by direct impact of raindrops. The litter also serves as a food and energy source for soil organisms, which encourages the formation of soil organic matter and faunal passages leading to high infiltration rates. Where forests are not protected from grazing and litter is consumed by livestock, removed for use as mulch as in parts of Nepal, or is lost in fires, the surface cover may be diminished to such an extent that the soil becomes bare.

This is likely to be more serious under trees that discourage the growth of understorey herbs and shrubs, such as teak (Tectona grandis) and some species of Eucalyptus due to shade, high water use - especially by Eucalyptus - and to a lesser extent because of the acid nature of the litter. If the tree canopy is high, accumulated rainwater drops that fall off the leaves may be larger than normal raindrops and can fall with sufficient velocity to cause more damage to the soil than if there were no tree cover. This can lead to a pronounced loss of soil porosity, as can trampling by livestock, resulting in restricted infiltration and high runoff despite the high canopy cover.

Infiltration by Detaining Runoff with Physical Structures

Alternative, but less favourable solutions to restricted infiltration are the use of physical structures, which may be necessary under certain situations:

- When it is not immediately feasible to implement conservation agriculture or simple soil cover because, for example, crop residues are used as fodder.
- As backup measures to support conservation agriculture where the problem of restricted infiltration is due to rainfall intensities that are higher than soil infiltration rates even in the presence of a residue cover.

In these situations, the volume of water soaking into the soil may be increased by giving more time for infiltration by slowing down runoff, by means of physical or vegetative structures constructed across the slope and parallel to the contour. Closely spaced structures on the contour (e.g. ridge and furrow series of planting lines and irregularities formed by contour tillage and crop management operations) may be formed over the whole field so that rainfall is detained where it falls. Widely spaced structures at intervals down the slope (e.g. fanya juu terraces, stone walls, earth bunds, live barriers and trash lines) used on their own without contour field operations between them will result in rainwater running downslope until it is detained or slowed down at the next barrier.

Contour Field Operations

On sloping land all field operations such as tillage, planting, weed control, spraying and harvesting should be carried out along the contour. Ridges and mini-depressions along the contour create small storage volumes where rainwater can accumulate, allowing more time for infiltration. Field operations conducted in a downslope direction can cause a devastating impact resulting in high runoff losses and soil erosion.

Narrowly spaced contour planting ridges with and without cross ties have the advantage of detaining rainwater where it falls so that there is more time for soak-in, and can be an effective means of encouraging infiltration and preventing runoff in semiarid and the drier subhumid areas. An additional advantage is that working along the contour makes operations such as harvesting easier and quicker.

Challenges of Contour Field Operations

The surface depressions have limited capacity to retain water and on sloping land the effective storage volume rapidly diminishes as slope increases. On slopes greater than 5 percent the effective storage volumes are considerably reduced. Reductions in storage volume will also occur on soils with a low structural stability, as the small ridges slump into depressions on becoming wet. Substantial runoff can occur even on land of 1-2 percent slope when the soils are of low stability and susceptible to crusting. Even on structurally stable soils, depressions may be quickly overtopped by the accumulation of rainwater from all but the lightest of rainstorms.

Factors for the adopting of contour field operations

The only exceptions to contour cultivation may be in high-rainfall areas where the soils have high infiltration rates and high susceptibilities to mass movements, e.g. landslides and mudflows. In these situations high soil water content increases the risk of mass movements, and so it may be better to encourage controlled runoff of some of the rainfall. Since the effectiveness of contour field operations in reducing runoff is limited on all but the gentlest slopes, it should be considered as just one of the practices necessary to increase water availability.

Contour Planting Ridges and Tied Ridges

In tied ridges the ties are constructed at intervals across the furrows formed by the contour ridges. These structures are usually constructed with animal traction or tractor power and may be formed annually or can be semi-permanent. They may also be made by hand but labour demands are high. The precise form and management of contour ridges and tied ridges vary considerably, with the optimum design and management being dependent on the crop, rainfall and soil type.

Contour ridges run the risk of being overtopped if too much rainwater accumulates within the furrows. They also may be breached or collapse at low points where large volumes of runoff accumulate from along the furrows. If large volumes of water frequently accumulate a subsurface pan or horizon of restricted permeability may be present beneath the furrows.

These risks can be reduced by carefully laying out and maintaining the ridges and furrows to ensure there are no low points and by constructing tied ridges to prevent lateral movement of water along the furrows towards any low points that may exist. The ties should be spaced at 1 to 3 metre intervals along the furrows and no more than half to two-thirds the height of the ridges. Although tied ridges require additional work, they provide good insurance against the collapse of ridges at low points during heavy rains and the loss of rainwater by discharge from the ends of the furrows if a slight gradient exists.

The furrows of contour ridges are normally aligned parallel to the contour. However, if very large volumes of runoff are periodically expected, tied ridges should be installed and the furrows constructed on a slight gradient (never steeper than 2 percent in the direction of a natural watercourse) so that excess rainwater is discharged along the furrows to prevent overtopping of the ridges. In these circumstances well-designed discharge points will be necessary at the furrow outlets. The size and spacing of the ridges should coincide with the crop's recommended spacing, furrow width and depth.

Ridges and tied ridges may be constructed prior to, or after, planting. Maize is often planted on the flat, and the ridges constructed at the time of the first weeding about 30 days after planting, which saves labour. Clearly, the earlier ridges are constructed the more rainwater they will be able to detain, and the greater the probability of a good yield. The time when ridges are constructed is also a convenient time to simultaneously incorporate manures.

Advantages of Contour Ridges and Tied Ridges

The main advantage of contour ridges and tied ridges is the greater accumulation of rainwater within the furrows due to the retention of potential runoff. The concentration of water in the furrows encourages deeper percolation, but for this to be useful to the crop, the soil's AWC must be sufficiently high to retain the accumulated water within reach of the crop roots. Sandy soils with a low AWC may permit a large proportion of the rainwater to drain beyond the zone penetrated by the roots.

Challenges of Narrowly Spaced Contour Ridges and Tied Ridges

The continual formation of ridges each year by hand or by mechanisation, combined with trampling along the furrows, may result in the formation of a compacted horizon at the base of the ridges, which can prevent roots from penetrating into deeper layers. This will counter the advantages provided by the ridges of increasing the supply of available water. The exposure of the soil surface leads to an accelerated loss of soil organic matter and surface crusting due to the effects of tillage, raindrop action and direct exposure to the sun, and very little macrofaunal activity. Consequently, the soils rapidly become degraded.

Another constraint is the time required to construct contour ridges, with even more time needed for tied ridges. The manual construction of contour ridges needs about 100 hours per hectare, and heavy textured soils will be even more demanding. To form ridges by hand or by animal traction in hardsetting soils will generally only be possible once the first rains have moistened the soil. The process of manually constructing contour ridges on sloping land, where the farmer faces uphill and pulls the soil into ridges with a hoe, causes soil to move downhill, so encouraging soil erosion.

Favouring Factors for Adopting Narrowly Spaced Contour Ridges and Tied Ridges

Contour ridges and tied ridges are most suited to areas suffering from water deficits where it is not feasible to provide a soil cover to enhance infiltration and reduce runoff through the use of crop residues, mulching materials or cover crops. The manual implementation of these structures will only be possible where sufficient labour is available and where farmers consider that the high labour requirement is justified by the value of the crop. These structures are particularly suitable for the production of tuber crops.

Impermeable and Permeable Contour Barriers at Discrete Intervals Downslope

These structures include stone lines, walls, earth banks, trash lines, live barriers and similar constructions. They have usually been installed to prevent small rills developing into gullies by limiting the area over which runoff collects, with or without sideways diversion into prepared waterways for safe disposal downslope. The barriers, which they provide may, if well maintained, accumulate soil which has been eroded from upslope.

In many situations, the chief benefit of laying out structures along the contour, at discrete intervals downslope, is their use as guidelines for the alignment of contour field operations in the cropping areas between them. The capture and soak-in of runoff along the upper sides of these structures may be considered as an added, rather than a primary, benefit.

The more closely spaced the banks, the more frequently runoff will be intercepted, but the more of the farmer's land will be taken out of production, unless some useful crop is planted along the earth bank. In semiarid areas structures can be designed for the purpose of water harvesting, which provides the extra water needed for adequate yields, if only from a relatively narrow strip. If the strip of land immediately upslope of the barrier has been made impermeable by passage of machinery or of feet, or the soil itself is relatively impermeable, temporary or more long-lasting local waterlogging can be induced.

If more runoff accumulates than the structure can hold back on its upslope side, it will overtop and may break, with the resulting concentrated flow of accumulated water often causing more damage downslope than if the structure had not been there at all. Conditions favourable for adoption of impermeable cross-slope barriers for water conservation

In areas of moderate to high rainfall, such barriers may be appropriate where they complement water-absorptive conditions, good surface cover and/or ridge and furrows, with or without tied ridges. If they are laid out on the level contour they may have some small additional effect on increasing water in the soil.

Permeable Cross-slope Barriers

Permeable barriers, which may be accumulations of stalks, branches, crop residues, leaves (trash lines) without or with a line of one or more crops, forage grasses, shrubs or trees (live barriers) may impede but not stop runoff. The lower speed as runoff passes tortuously through the material provides an opportunity for infiltration. The live barrier may benefit from the additional soil moisture, but the additional transpiration through deep-rooted plants may minimise the volume, which could flow beyond the roots to groundwater. If the farmer receives no benefit from the live barriers, then the competition effects for light and moisture would be a disincentive.

Bench-type Terraces

These structures are a total modification of natural land slopes into a series of platforms which are almost level or slope at shallow gradient across or along the terrace. Controlling the gradient in this way allows management of water movement on what were formerly steep slopes. The cultivation platform may be continuous along the slope or discontinuous (orchard or platform terrace). The surface of the intervening uncontrolled slopes is preferably covered with a close-growing grass or legume as a soil-protecting cover.

Bench-type terraces are arduous and expensive to construct, requiring up to 700 person-days per hectare. Their capacity to receive and store rainwater depends on the depth, condition and quality of the soil into which they have been constructed. In semiarid areas they may be able to catch and detain all the rain that falls. In places with greater volume and frequency of rainfall, provision may have to be made for disposal of excess water down very steep waterways, and there is also an added danger of landslips if the benches become saturated.

Deep Tillage to Increase Subsoil Porosity and Permeability

Rainwater infiltration may be restricted in soils where the pore spaces rapidly become saturated with water because of the presence of dense subsoil horizons of low permeability. In these situations an initial deep tillage of the whole field with a tined implement, subsoiler or paraplow to break-up the dense horizon may improve subsoil permeability and so allow more rainwater to infiltrate. By improving subsoil permeability the rate of oxygen supply to the crop roots will also improve. However, the beneficial effects of deep tillage may only last 2 to 3 years.

Evaporation and Excessive Transpiration Losses

The most effective solution to high evaporation losses of soil water is a cover of plant residues on the soil surface. Agronomic practices that increase shading of the soil surface, and physical structures that concentrate rainwater, encouraging percolation to deeper layers, also reduce evaporation losses. Wasteful transpiration losses may be the result of weeds or excessive crop transpiration in hot windy conditions, and can be reduced by appropriate weed control practices and windbreaks, respectively.

Reducing Evaporation from the Soil Surface

Surface residues reduce soil water losses through evaporation by acting as an insulating layer. This diminishes the temperature of the surface soil and eliminates the effect of wind. Heat from the sun is only slowly transmitted from the surface of the residues through the air trapped within the layer of residues to the soil surface. Consequently the soil surface remains cooler and the rate of evaporation of soil water is slowed down. The thicker the layer of trapped air, the greater will be the insulating effect, and the quantity of residues required to reduce evaporation losses is considerably greater than the quantity needed to ensure that most rainfall infiltrates where it falls.

For example, in Uganda, farmers traditionally apply between 8 and 40 t/ha of mulch to bananas, whereas 4-5 t/ha are probably sufficient to minimise runoff and allow most of the rainfall to infiltrate. Banana yields respond very favourably to mulching, and applying 5-10 cm depth of maize stover and Paspalum sp. to bananas at Sendusu, Uganda, increased yields from 4.3 to 10.8 t/ha.

Experiments in Uganda have shown that yields approximately doubled when 30-40 t/ha of mulch were applied compared with applications of 10-20 t/ha. This yield increase was mainly attributed to lowered evaporation losses. Protecting the soil surface from wind also slows down evaporation by reducing the rate at which water vapour is removed from the soil surface.

The use of residue covers for conserving soil moisture in the topsoil and increasing yields is particularly important in regions with limited rainfall and high evaporation rates. It is also important for shallow-rooted crops, e.g. bananas, tea, coffee, pineapple, and vegetables such as onion, lettuce, cabbage and carrots. Residue cover can also be very beneficial in reducing water losses by evaporation from soils with a shallow water table (less than 1 to 2 metres), from which there may be capillary rise of the subsurface water. Such soils are often used in horticultural production.

However, the main disadvantage of using residue covers for reducing direct evaporation is the large quantities of residues required to significantly reduce evaporation. Often, the regions with high evaporation losses also suffer from a shortage of rainfall, which restricts the production of vegetative matter. Frequently there are also other demands on residues, which take priority such as fodder, thatching and construction.

Reducing Excessive Transpiration

In hot windy weather, the rate of loss of water through plants by transpiration can be very high and can result in early depletion of limited soil moisture reserves. This in turn can lead to serious water stresses developing in plants - both crops and weeds - before their cycle of growth to maturity has been completed.

Weed Control

Loss of soil water through weed transpiration can seriously reduce the amount of water available to crops. Consequently, timely and effective weed control practices are essential. The presence of a thick layer of residues on the surface is a very effective way of controlling weeds. Where weed control measures are needed, the use of herbicides or appropriate crop rotations is often preferable from a conservationist perspective to mechanical weed control, unless it is practised with no soil disturbance. Post-emergence herbicides leave weed residues on the soil surface as a protective cover whereas cultivation leaves soil exposed to the impact of raindrops and sun, accelerates drying of the surface soil and tends to disrupt and destroy soil porosity through smearing and compaction.

Windbreaks

When crops are exposed to strong winds in a dry environment the water that has been transpired by the crop is rapidly removed from the leaf surfaces into the atmosphere. This encourages a more rapid movement of water up through the crop and much greater absorption of water from the soil. Strong winds can therefore cause excessive crop transpiration rates and an unnecessary loss of soil water.

Windbreaks will significantly reduce wind speed and so reduce crop transpiration rates and the unnecessary loss of soil water. Windbreaks are usually established by planting single, double or triple rows of trees, but sugar cane or tall grass species may also be used. In areas where forests are being cleared for agricultural development, strips of the original forest may be left as natural windbreaks.

Important considerations in the design of planted windbreaks are their composition, orientation, height, porosity and spacing. Windbreaks should be oriented at right angles to the direction of the prevailing winds during the growing season. As a general rule, they should occupy no more than 5 percent of the cropped area. For small production units a single row of trees is usually most appropriate. Paths and roads should not cross windbreaks to avoid channelling of the wind through the openings at high velocities. The tree species selected should be adapted to the climate and soils of the area.

The foliage should not be so dense that most of the wind is forced to pass over the top of the windbreak, as this will cause severe turbulence on the downwind side of the windbreak, which can seriously damage the crop. The porosity of the windbreak vegetation should ideally be 40 percent so that part of the wind passes through the windbreak. This will give a 50 percent reduction in the velocity of the wind within a distance of ten times the height of the trees.

Maintenance of the windbreaks is important to ensure that no holes appear, to regulate the porosity of the vegetation to wind and to avoid excessive shading and weed infestation of adjacent crops. Natural windbreaks are strips of forest left after deforestation. Since a much drier and windier microclimate develops in these strips of forest compared with that in the undisturbed forest, many trees in natural windbreaks often die, sometimes leaving holes through which the wind passes at increased velocity. The important guideline for natural windbreaks, as for planted windbreaks, is that the porosity of the vegetation should be about 40 percent.

In open forests in particular, natural windbreaks may need to be substantially wider than planted windbreaks to allow for the death of some trees. Alternatively, planting individual trees to fill gaps, or enriching the natural windbreak with one or two rows of additional trees may be necessary to produce a protective cover of 40 percent porosity. Well-designed windbreaks will significantly reduce evapotranspiration rates of crops in windy conditions resulting in the conservation of soil water and less subsequent moisture stress when water is limiting.

A 50 percent reduction in wind velocity (from 32 to 16 km/h) will reduce evapotranspiration rates by 33 percent. Windbreaks may provide additional benefits to crops by reducing mechanical damage and the loss of flowers, and by creating better conditions for insect pollination. They are also beneficial in reducing wind erosion, especially in fine-sandy and silty soils, and in diminishing air pollution problems. Depending on the tree species selected, windbreaks may also provide fruit, nuts, fodder and timber, but the harvesting of these products must not result in pronounced gaps being formed within the windbreak.

The main disadvantage for farmers with small plots is the loss of cropping area due to the windbreak and the risks of competition between the windbreak and the crop for water, nutrients and light leading to lower crop yields. This zone of competition may extend over a distance equal to 1.5 times the height of the windbreak. In areas where there are severe shortages of fodder, fuelwood and timber, windbreaks may need to be fenced to prevent indiscriminate grazing and harvesting. To ensure that wind cannot pass around the ends of individual windbreaks, the establishment of windbreaks should be planned on a community basis.

Factors for Adopting Windbreaks

Windbreaks will be favoured in areas subject to strong dry winds during the growing season, and where windbreaks cause a net gain in soil water (i.e. where the gain in soil water due to reduced crop transpiration exceeds the loss of water due to windbreak transpiration). Windbreaks are also likely to be favoured where they consist of species that provide additional benefits, such as fodder, fruit, nuts, fuelwood and timber that can be harvested without damaging the windbreak.

Shade

Shade can be provided by all manner of materials, whether artificial such as nets, cloths, plastic sheets and others, or plant-derived, such as cut branches, cut grass supported on nets, or living trees which provide high-level and wide-spreading shade. Shade is necessary in plant nurseries in hot regions to protect seedlings and other plants with shallow roots from rapid desiccation. While shade may ameliorate the severity of hot dry conditions and limit undesirable losses of soil moisture, it can also be so dense as to limit solar energy reaching leaf surfaces and limit photosynthesis and growth rates.

Where shade may be desirable, its density should be adjusted to provide an appropriate balance between losing water too fast, limiting sunlight intensity and avoiding scorching of leaves due to temporary dehydration and cell-damaging high temperatures. Using living shrubs and trees to provide long-term shade for tea and coffee can cause difficulties in maintaining the desired degree of shade above the crop over the long term.

Reducing Rainwater Drainage Zones

Soils without Restricted Rooting

In regions where much of the rainfall occurs as light showers, the concentration of rainwater as near as possible to the crop will cause more of the rainwater to infiltrate deeply, where it is less susceptible to evaporation. In order not to lose this water by drainage beyond the crop's rooting zone and where there is no rooting restriction some solutions can be adapted, such as increasing the capacity of soils to retain water within the rooting zone, early planting to accelerate root development or changing to deeper-rooting crops.

Increasing Water Capacity of Soil

The addition of large quantities of organic manure will increase the available water capacity (AWC) of soils and in theory this is a useful practice for reducing deep drainage losses. However, even in temperate climates the quantities of organic materials required to markedly increase AWC are very high, applications must be continued over many years and usually affect only the plough-layer depth. In tropical zones, where organic matter decomposition rates are much higher, the influence of organic manures on AWC is likely to be even less. Nevertheless, this practice may be feasible for small-scale farmers growing high-value crops where large quantities of organic manures and labour are readily available.

Dry Planting

In low rainfall areas, it is frequently difficult to know when the rains have truly started, as initial rains are often followed by a dry period. Many farmers wait until the topsoil has been moistened to a depth of about 15-20 cm before planting, so that even if there is a subsequent short dry period there is sufficient water within the soil. However, this results in a delay in planting and for every day's delay yields will decrease, largely due to the loss of rainwater by drainage and evaporation, together with the loss of some released nutrients.

To overcome this problem and to allow crops to develop deeper rooting systems earlier on so that more of the rainfall can be utilised during the initial stages of the season, some farmers "dry plant" when soils are dry prior to the onset of the rains. To avoid premature germination before sufficient rain has fallen, the seeds are usually placed deeper than normal. Dry planting also has the advantage of spreading labour over a longer period. Crops may also benefit from this practice by being able to utilise the nitrogen released at the start of the rains from the decomposition of soil organic matter, which reduces leaching and pollution of groundwater. However, there are a number of problems associated with dry planting, notably that some soils, and in particular hardsetting soils, are difficult if not impossible to till when dry. If seeds are not planted sufficiently deeply, they may germinate at the first rains and then die during a subsequent dry period.

Early Root Development for Plant Nutrition

Applying fertiliser to speed up crop canopy development and increase the shading of the soil surface will decrease the soil water lost by evaporation so that more is available to the crop. Planting crops equidistantly (i.e. with between-row spacing similar to within-row spacing) so that the soil surface becomes shaded more quickly would also be expected to reduce the proportion of soil water lost by evaporation. However, the effects of these agronomic practices on reducing evaporation losses will be much less than applying surface residues.

Introducing Deep-rooting Crops

On permeable sandy soils that retain small quantities of available water for crop use, it is preferable to introduce deep-rooting crops that can utilise soil water at depth that would not be available to shallow-rooting crops. Examples of deep-rooting crops are almond, barley, cassava, citrus, cotton, grape, groundnut, olive, pearl millet, pigeon pea, safflower, sisal, sorghum, sunflower, sweet potato and wheat.

Restricted Rooting

The type of solution to be applied will depend on the cause of root restriction. The most frequent cause is physical root restriction due to a lack of pores that are large enough to be readily penetrated by roots or which can be sufficiently widened by the growing roots. This condition occurs in dense layers, such as plough pans formed by tillage, but also in naturally occurring dense layers as found in hardsetting soils. Root restriction may be overcome, at least temporarily, by biological or mechanical means. In addition to eradicating the causes of root restriction it is also important to take steps to avoid future recurrence of the problem by, for example, introducing conservation agriculture where dense layers have been formed by tillage.

Less common causes of restricted rooting are chemical restrictions due to the presence of toxic concentrations of aluminium or manganese, high salinity or severe nutrient deficiencies, especially of phosphorus. A lack of oxygen due to a fluctuating water table may also restrict root development. While the water table is high, root development for most crops will be restricted to the soil immediately above the upper level of the water table but the crop will not suffer from a lack of moisture. If the water table then falls relatively quickly to a substantially lower level, for example at flowering, when the crop has still to reach physiological maturity but the roots have ceased growing, the roots may be left stranded in the dry soil without access to the moisture in deeper layers.

The causes of restricted rooting given above can, where appropriate, be overcome by the application of lime, or lime and the more mobile gypsum, to eradicate aluminium and manganese toxicities; leaching to reduce salinity hazards; fertilisers to rectify nutrient deficiencies; or drainage to remedy the lack of oxygen from a fluctuating water table.

The principal biological method of restoring the porosity of root-restricting layers is to place the land in fallow and utilise the roots of natural vegetation or planted cover crops to act as biological subsoilers penetrating the dense root-restricting horizons. The stability of root channels created by plant roots will be greater than that of channels formed by mechanical methods because of the release of organic substances from the roots that stabilise the channel surfaces. Once the roots have died and shrunk, these pores will be sufficiently large and stable to enable the roots of subsequent crops to penetrate.

Land may be left in fallow for 2-3 years for natural bush or forest vegetation to regenerate. Alternatively, planting selected species that are effective in regenerating soil structure can enrich the natural fallow. A cover crop may be sown to serve as a planted fallow. Promising cover crop species that have been shown to have potential as biological subsoilers are the grasses Bahia grass (Paspalum notatum), Festuca elatior, Guinea grass, and alfalfa , pigeon pea (Cajanus cajan) and cowpea (Vigna unguiculata).

Biological methods are generally much cheaper to implement and their benefits are longer-lasting than mechanical methods. An important advantage of vegetative fallows is that they greatly improve the physical, chemical and biological fertility of the soil due to the large quantities of organic matter produced and added to the soil. Tree fallows can be beneficial in supplying fuelwood, construction materials and other products, provided the harvesting of these materials does not reduce the beneficial effects of the fallow on soil chemical fertility.

Factors for Adopting Biological Methods

The use of natural biological methods will be favoured by farmers who have sufficient land. They can take some of it out of production and place it into fallow while the slow process of natural regeneration of soil porosity takes place. The use of cover crop fallows is often a rapid process which enables land to be more quickly returned to production. Natural fallows in which there is a regeneration of tree vegetation are more likely to be adopted by farmers who wish to change the land use of the recuperated area to forest or perennial tree crops.

Mechanical Methods to Physical Root Restriction

The aim of mechanical methods is to break-up the compacted or naturally dense root-restricting layer in order to create larger pores through which crop roots can penetrate. This is accomplished by the implement slightly lifting and breaking the compacted or dense layer. The operation may be carried out over the whole of the field, or merely along the rows where the crop is to be planted. The latter, known as in-row subsoiling, is much quicker and requires less draught power, but the crop must be sown with precision directly over the loosened rows.

Mechanical Disruption of Shallow Root-restricting Layers

Shallow root-restricting layers such as hoe pans are typically produced at 5 to 8 cm depth, and the easiest means of breaking them up are with ox-drawn rippers or tractor-mounted chisel ploughs. Most farmers relying on manual tillage will probably have to use hand tools to break the hoe pans by methods such as double digging, which are very labour-intensive. To break-up compacted layers in the dry season when the soil is very hard may require robust tools different from those the farmer normally uses for tillage, such as pickaxe, mattock, three-tined hoe (jembe) or a long crowbar.

Mechanical Disruption of Moderately Deep Root-restricting Layers

Deeper root-restricting layers such as plough pans are formed at the lower limit to which the soil is tilled, and usually occur within the upper 10 to 25 cm of the soil profile. Plough pans formed by ox-drawn implements can usually be broken up using two passes of an ox-drawn ripper, whereas those formed by tractor-drawn or -mounted implements usually require a tractor-mounted subsoiler or paraplow.

Paraplows are similar to subsoilers except that the shanks are slanted sideways to the direction of travel, which enables soil to flow over the shanks. They are preferable to subsoilers as they bring fewer subsoil clods to the surface, require less draught power and cause less incorporation of surface residues that should ideally be left on the surface. Disc ploughs are less suitable because they invert the soil, incorporate most of the crop and weed residues and bring subsoil clods to the surface, resulting in the need for additional tillage.

If the root-restricting layer is to be disintegrated over the whole field, then as a rule of thumb the subsoiler or paraplow should penetrate to 1.5 times the depth to the lower limit of the root-restricting layer, and the spacing of the shanks should not be greater than this value. For example, if the root-restricting layer occurs at 10-24 cm depth, the shanks of the subsoiler or paraplow should penetrate to 36 cm and the spacing between the shanks should be no more than 36 cm. If the shanks are more widely spaced, there is a likelihood that the root-restricting layer will not be fully disrupted in the region midway between where the shanks passed. To avoid compaction from the wheels of the tractor, shanks should be positioned immediately behind the tractor's wheels. For in-row subsoiling, the shanks need only penetrate to the lower limit of the root-restricting layer, and the shank spacing should coincide with the planned row spacing of the crop.

Subsoiling should be carried out perpendicular to the normal direction of tillage and the soil should be dry to the depth of subsoiling to obtain good shattering. If the soil is moist or wet, there will be no shattering, merely the formation of channels gouged out where the subsoiler's points have passed.

Mechanical Disruption of Deep Root-restricting Layers

Subsoil compaction at 40 cm depth and greater is caused by the passage of very heavy equipment with high axle loads of at least 6 tonnes, such as combine harvesters and lorries laden with grain. At this depth the use of conventional subsoilers to loosen deep compacted layers is difficult and expensive because of the very high traction power needed. Vibratory and rocking subsoilers, in which the subsoiler points vibrate or rock using the tractor's power takeoff can work to 80 cm. depth, but require 75-100 HP. New implements have been developed employing elliptically moving blades or rotary hoes, which utilise a break-off-loosening mechanism to disintegrate compacted layers. They can be used to depths of 60 to 120 cm and at higher soil moisture contents than conventional subsoilers, but are very expensive and require high traction power.

The shattering and lifting of root-restricting layers by mechanical means creates larger pore spaces through which roots can penetrate, enabling them to reach and take advantage of soil moisture and nutrients stored in deeper layers. Consequently, crops are able to make more efficient use of the rainfall. The main effect of subsoiling is usually that of promoting deeper root growth, but if the root-restricting layers are so dense that rainwater movement is also limited, subsoiling may also facilitate the percolation of rainwater into deeper layers.

The development of improved rooting frequently increases crop and pasture yields. In Babati District, Tanzania, breaking up hardpans by subsoiling has almost tripled maize yields and quadrupled maize dry matter production. Increased yields from subsoiling are most likely in areas where yields are limited by rainfall, and the drier the season the greater the probable response to subsoiling.

In-row subsoiling, especially when it is combined with planting in a single operation, is particularly beneficial for hardsetting soils that rapidly form root-restricting layers on drying after being saturated with rain. This technique is most likely to be successful when associated with precision planting and controlled traffic, in which the passage of all machinery wheels is restricted to permanent tracks. The benefits of subsoiling are likely to be greatest when immediately followed by the establishment of a dense cover crop with a strong rooting system that helps stabilise the new pore spaces created. The cover crop should then be followed by a system of conservation agriculture in which the absence of tillage reduces the recurrence of further compaction.

The principal disadvantage of mechanically breaking up root-restricting soil layers is the high power requirement, whether it is manual, animal or mechanical. Since most farmers do not have access to more than that which they use for land preparation, the process is inevitably slow. Some soils become so extremely hard during the dry season, that the farmer's normal draught power is incapable of penetrating the soil in order to break-up the root-restricting layer. It is then necessary to wait for the beginning of the rains to moisten and soften the soil before it becomes possible to break-up the compacted layer, but this may coincide with the critical time of land preparation and planting.

This problem can apply equally to farmers using animal traction or tractors and to those using hand tools as their source of power. Subsoiling operations are ineffective when the dense or compacted layers are wet or very moist as no shattering effect takes place. Farmers often lack the necessary implements, whether they are pickaxes for farmers relying on manual power, rippers for animal traction farmers, or subsoilers or paraplows for mechanised farmers. The use of normal land preparation implements will generally be less satisfactory.

For example, disc ploughs can be used to break-up plough pans, but they invert the soil bringing large clods of subsoil to the surface and form an uneven surface that needs additional tillage to create a seed bed. Disc ploughs also incorporate the residues of crops and weeds, when ideally they should be left as a protective layer on the soil surface. Repeated use of disc equipment, especially heavy-duty disc harrows, can produce an almost impermeable compacted pan in only a few seasons. These pans have been the cause of increasingly severe runoff and erosion from millions of hectares in Brazil, before the use of disc equipment was abandoned in favour of minimum tillage with tines, and subsequently by no-till systems. When bulky crop residues are left on the surface, especially the stiff residues of maize, sorghum and cotton, the performance of subsoilers and paraplows is considerably impaired unless they are fitted with front cutting discs.

If subsoiling is followed by conventional tillage, the beneficial effects are only likely to persist for 2 or possibly 3 years and so the subsoiling has to be regularly repeated. The speed with which the root-restricting layer reform will depend on the number of tillage and other field operations, the moisture content of the soil at the time of these operations and the susceptibility of the soil to compaction. Fine-sandy and silty soils and those with impeded drainage are most susceptible to compaction.

To improve the physical conditions of hardsetting soils requires the incorporation of large quantities of organic material into the dense layers and the regeneration process is likely to be slow. For hardsetting soils, in-row subsoiling may be necessary each year. These disadvantages can be overcome by adopting reduced tillage, or preferably zero tillage as in conservation agriculture, or by controlled traffic in which all machinery follows the same tracks year after year, leaving the cropped strips untouched. Thorough loosening of soils by subsoiling may render them more susceptible to compaction if they are subsequently subjected to high pressures, as from excessive tillage or the passage of very heavy machinery. The recompaction may be worse than the original state of compaction.

Subsoiling heavy textured soils, such as vertisols, can greatly increase the quantity of rainwater that reaches the subsoil, resulting in a marked reduction in the soil's bearing strength, i.e. its capacity to support heavy machinery. It should be noted that subsoiling to any given depth produces a high proportion of very large soil pores and fissures, a situation favouring better penetration of roots and of rainwater. It will not however produce any significant increase in the range of smaller soil pores, which make up the water-retention capacity of the soil.

Favouring Factors for Adopting Mechanical Methods

The adoption of mechanical methods to overcome physical root restriction will be favoured where yields are frequently limited by low rainfall. Under such conditions it becomes important that as much of the rainfall as possible is stored within the soil profile, and that the crop's roots have access to all of the stored soil moisture. Mechanical methods will be favoured where farmers have access to tractors and subsoilers or paraplows, and where land cannot be taken out of production and put down to fallow for 2 to 3 years.

Chemical Solutions to Restricted Root Growth

Root development is sometimes restricted by unfavourable soil chemical conditions, such as severe nutrient deficiencies, aluminium or manganese toxicity and salinity. The nutrient which most commonly restricts root development is phosphorus and the application of P fertilisers to phosphorus-deficient soils frequently encourages deeper rooting, enabling the crop to access more soil moisture and so increase productivity. The incorporation of lime without or with gypsum will reduce toxic concentrations of aluminium and/or manganese to non-toxic levels and so encourage deeper rooting. The greater solubility of gypsum compared with lime makes the former more suited to soils with aluminium or manganese toxicity problems in the subsoil, whereas the slowly soluble lime is most effective in topsoils. When high salt concentrations inhibit root development in irrigated soils, excess quantities of water should be applied sufficient to leach the salts out of the crop's rooting zone.

Usefulness of Low and Erratic Rainfall

Several approaches may be used to diminish the impact of low and erratic rainfall, viz. match land use to soil characteristics; use drought-resisting and drought-escaping crops; increase the efficiency with which crops utilise rainwater; concentrate rainfall by water harvesting; divert river water; intercept floodwater; and apply supplementary irrigation.

Match Land Use

Matching land use to the most suitable soil types within a farm may increase the efficiency with which the available soil water in the different soil types is utilised for crop production. Crop water requirements vary, as do the capacities of soils to retain and supply water to crops. Moreover, the variations in available water capacities (AWC) of soils often occur over short distances. Soils with high AWC will be expected to suffer less water loss from deep drainage and possibly from runoff. Consequently, greater quantities of rainwater will remain in the soil and so the potential crop-growing season will be longer assuming an adequate amount, distribution and infiltration of rainfall.

Matching crops with weak root systems, such as beans, to soils lacking root-impeding layers, would be expected to increase crop water use efficiency. Beans are more suited to freshly tilled soils, or to mature no-tilled soils where large numbers of channels suitable for root penetration have been created through the decomposition of old roots and soil faunal activities.

Allocating land use to suitable soil types may enable production to be intensified, leading to benefits in addition to that of higher water use efficiency. Thus, intensifying subsistence food crop production may liberate land for producing cash crops. Alternatively, it may allow land previously used for inappropriate, extensive and degrading forms of land use, to revert to natural vegetation, thereby reducing land degradation.

Drought-resistant and Drought-escaping Crops

Some crops can tolerate drought because they are able to resist a shortage of water, i.e. they are said to be drought-resistant. This is because either:

- — they can stop growing when water is unavailable by becoming dormant. When rain occurs they resume growing and developing as though nothing had happened; or
- — they have deep rooting systems, such as pigeon pea, and can absorb water from deep within the soil. This is important where occasional rainstorms wet the soil to great depth followed by long dry periods.

Pineapples and sisal resist the effects of drought due to their thick leaves that slow down water loss by transpiration. These crops, as well as sorghum, pearl millet, pigeon pea, cassava, groundnut and cowpea, are drought-resistant and suited to climates with a defined mid-season drought.

A drawback of drought-escaping crops is that their short growing season restricts yields compared with long-season cultivars, although under dry conditions they will outyield the long-season cultivars. For example, improved pearl millet varieties in Tanzania, which mature two weeks earlier than farmers' local varieties, have yielded 43 percent more (2.31 t/ha) than local varieties (1.62 t/ha). Applying fertilisers to counteract nutrient deficiencies can speed up crop maturity and so enable them to escape droughts more easily.

Drought-escaping crops are more suited to short rainy seasons, or to soils which can only store a limited quantity of water. It is therefore important to select drought-escaping crops and varieties whose maturation period matches the expected length of growing season. If possible very determinate varieties should be avoided so that the risks of the whole crop being adversely affected by dry periods is reduced. Unfortunately, farmers often do not have access to varieties that match the expected length of growing season, and the length of growing season may vary widely from year to year. It must be borne in mind that the choice of crops and varieties depends not only on their ability to resist or escape droughts, but also on their susceptibility to pests and diseases, labour requirements, availability of seed, ease of grain processing (threshing, dehulling and grinding), fuel requirements for cooking, and palatability.

Efficiency of Crop Water Use

Crop water use efficiency refers to the amount of dry matter produced for each millimetre of water that is transpired by the crop or evaporated by the soil, i.e. for each millimetre of evapotranspiration. Clearly, in dry areas the more efficient use the crop can make of the rainfall that infiltrates (referred to as the effective rainfall), the higher will be the yield. The following management practices influence crop water use efficiency:

Selecting Water-efficient Crops

A group of crops referred to as C4 crops, which include maize, sugar cane, sorghum and pearl millet, are physiologically much more efficient at producing dry matter for each millimetre of transpired water than other crops, referred to as C3 crops. But this distinction is most important in situations where rainfall is adequate. For areas where water deficits are common, the use of drought-resistant and drought-escaping crops is much more important.

Adjusting Plant Population to Expected Rainfall

A high plant population will use large amounts of water for transpiration during early growth provided sufficient water is available in the soil. Because of rapid shading of the soil by the crop foliage, less water will be lost by direct evaporation, ensuring a higher water use efficiency compared with low plant populations. High plant populations, and especially those with a more square planting arrangement, also increase water use efficiency through the quicker development of cover and therefore less weed growth.

Although evaporation losses are greater for low plant populations, soil texture and the frequency of rainfall events also influence the amount of water lost. Sandy soils in areas where rainfall occurs in few heavy storms will suffer less evaporation than medium or fine textured soils in areas with frequent rainfall events.

Where rainfall is erratic the situation is complicated, and becomes more than just a matter of water use efficiency. Farmers then face the dilemma of whether to sow at a low density to ensure some yield in bad years but underperforming in good years, or to use a high population to maximise yields in good rainfall years but to harvest very little, if anything, in bad years. If farmers have sufficient land they can opt for both options, i.e. an area with low population and another area with high population, but many small-scale farmers possess insufficient land for this to be feasible.

Response farming is an approach for matching crop management to estimated seasonal rainfall in variable rainfall zones. Plant populations and N fertiliser applications are adjusted after the crop has been established on the basis of information about the expected rainfall. Initially, the crop is sown at a high population assuming a good rainfall season, and with a low application of N fertiliser. The expected potential of the season (good, fair, poor) is determined on the basis of the anticipated amount of rainfall during the first 30-50 days, derived from as many years' records as are available. Decisions are then made according to the amount of rainfall early in the season on whether or not to thin or to apply additional N fertiliser. So far, this practice has not been adopted by farmers because of the great variations in seasonal rainfall over short distances, because farmers usually intercrop, and because of the initial wastage of water that occurs if crops are thinned after 30 or more days.

8

ICTs for Improving Crop Management

Most farmers tend to follow recommended practices regarding field operations and input levels, with relatively minor modifications to suit local conditions especially soils and climate. The conventional production system is a relatively high input: high output system, whereby mechanised field operations, hybrid varieties, and chemical additives and controls are used in combinations which, subject to the uncertainty of weather conditions, achieve reasonably consistent yield performance year to year.

Relative to other business activities, the manager of a farm contends not only with the risks of the market place, but also with the uncertainties of the production system, given its reliance on natural and biological processes. There are a large number of external, non-controllable factors, which influence the technical and economic performance of the farm system, such as prevailing weather conditions or the incidence of pests and diseases.

Decisions on cropping practices are required "upfront" without full knowledge of future events and resultant outcomes. Furthermore, the response of the system to actions taken cannot be predicted for sure, and decisions made at one point in the production cycle have considerable, albeit uncertain, influences on later outcomes and decisions. Farmers tend to be risk averters, adopting risk minimising or satisfying strategies which have a reasonable chance of reliable performance across the range of possible eventualities. This is evident in practice.

Farmers tend to concentrate on crops for which markets and prices are relatively stable or guaranteed. They may negotiate contracts to supply against agreed forward prices. They may also adopt methods of production such as chemical application levels, which are likely to achieve minimum acceptable, rather than maximum value-added, especially when the latter runs the risk of failure.

Conventional farming practices have evolved from a combination of experience and scientific knowledge. They are modified according to local circumstances such as soils and climate, but for the most part represent prescriptions for achieving the financial sustainability of the arable farm business. High yields and output are perceived to be the key to success. This is best achieved by relatively high variable inputs such as chemicals, which give relatively average returns per unit applied (even though in many years or in many parts of the farm, marginal returns may be low or negligible). In this respect, an element of the expenditure in the conventional package is an insurance premium, which accommodates the lack of knowledge about the future or the inherent variability of the production system.

For instance, in many years the capacity of the machinery inventory or the level of pesticide application may be greater than necessary. In any one year the variation in crop condition or soil type across the farm may not warrant uniform treatment. In an uncertain business environment, however, over capacity or blanket prescriptions are often preferred to taking risks, even though, over a period of years, lower cost, risky strategies may give better average financial returns. Furthermore, the conventional package reduces the demands on farm management expertise.

Risky systems are much more difficult to manage. Conventional methods economise on management skill. Alternative farming strategies, such as low input: low output, extensification, and integrated crop management systems have been tried and indeed promoted as financially attractive options. The results from such trials show that outputs can be similar to, and inputs significantly lower, than conventional systems.

However, such analysis misses the point. For the majority of farmers, these options are not attractive for widespread adoption because they are inherently much more risky and troublesome than the conventional package. The conventional system is a surer bet.

Role of IT in Farm Management

In the context of farm management, IT involves the use of digital information and control systems which acquire and use information to reduce risk, improve the efficiency of decision making, and assist with the implementation and evaluation of improved business management strategies. IT improves the knowledge base and increases the ability to control the production system in order to reduce risk and uncertainty, better identify the extent of variation in the system and thereby formulate strategies which explicitly address this variation, and supports the processes to implement these strategies. In this respect IT is essentially concerned with risk management.

From a management viewpoint, IT can be classified into two broad categories:

1. Information management and decision support:
   — Methods for the collection, assembly, logging and retrieval of data
   — Methods for data analysis, interpretation and decision making
2. Implementation: monitoring and control systems

New Trends

Since the early 1980s, there has been increasing concern that agricultural practice has caused damage to 'the environment', in a range of ways. These include:

— Possible deleterious effects on the farmland ecosystem itself, which may reduce the sustainability of agricultural practice

— Reductions or changes in desirable non-farmed habitats within rural areas,

— Direct effects on human health

Social effects, such as the non-renewable-resource costs of current agricultural practice, unpopular changes in landscape appearance, in rural employment and social structure. Predicting the effects of changes in management which are likely to arise from the adoption of a 'high information' strategy is only possible in very general terms, although some attempts at quantification of new strategies have been made in the USA. Most of the research to date has concentrated on quantifying the effects of historic changes in practice, which, until at least the mid 1980s, usually involved increasing use of fertilisers, biologically active inputs, and capital. It would not be appropriate simply to assume that the results of a reduction in inputs would be the exact reverse of the effects of increasing inputs.

There are now limited data from experiments which deliberately look at such reductions but these are insufficient to provide quantitative estimates of effects on a wide scale. Data from organic farming systems are more widely available, but represent an extreme form of management that is unlikely to be reflected in majority practice. Similarly, data from the initial years of set aside may not provide a reliable basis for estimating the effects of selective reductions in use of inputs within active husbandry systems.

A basic comparison will be made between 'conventional' and a 'high information' strategy, which assumes that much more detailed management information is available, so that 'what-if' investigations can be made readily. Within such a system, the final decisions made will vary, depending on the values of the decision maker, but the opportunity will exist to take into account environmental as well as economic management variables. The weighting put on environmental variables will depend on the decision maker, and may well range from zero to relatively high rankings, so the actual implementation arising from a 'high information' strategy will also vary.

However, it is assumed that in general, the total use of biologically active inputs will be reduced in the 'high information' strategy. In the arable and grassland context, traditional managers have tried to treat fields uniformly. All the field operations have been managed to produce an even distribution, even though it is known that the conditions may vary across the field. High or low yielding areas may have been identified but only on the small traditional farms was there the capability to treat selectively.

In recent years the problem has become worse as arable fields have increased in size due to economies of scale that were gained from the mechanisation sector. Mechanisation is no longer the limiting factor to efficient production but has now become the arable inputs such as seed, fertiliser, sprays, etc. From initial studies in Europe and the USA it has been found that there are significant potential benefits from selective treatment, in both the economic and environmental senses. If the spatial variability of the land and crops are investigated, then the potential savings become apparent. The higher the variability, the higher the potential savings.

The long term trend facing farming in Northern Europe is that of a continuing deterioration in the terms of trade whereby the prices of agricultural outputs fall relative to those of inputs. The response by farmers in the past has been a mixture of intensification and amalgamation, protecting incomes by yield improvement through high use of variable inputs such as fertiliser, and economising on fixed inputs such as labour and machinery by enlarging the scale of operation. Given the financial pressures on farmers, improved farm business management, supported by IT, becomes a more critical factor in ensuring financial survival.

More recently, farmers have found it beneficial to target their output to the changing needs of the food market, especially regarding quality and supply assurance. The liberalisation of agricultural production and marketing will hasten this process. These changes will further encourage the use of IT as part of a sophisticated production and marketing system. Pressures to protect incomes is likely to encourage the use of IT as a means of achieving more economic use of inputs as well as enabling production and marketing activities which increase the "added-value" at farm level.

The use of IT, particularly for business management, is likely to become an integral part of commercial farming systems and a pre-requisite for successful management. To this extent, IT will encourage its own diffusion amongst farmers. Those who do not use it will be disadvantaged.

In addition to economic pressure, there are likely to be greater constraints on land use and farming practice in order to meet environmental criteria. These constraints are emerging in the form of policy initiatives such as the Nitrogen and Environmental Sensitive Area Schemes and the EU Nitrification Directive. For the most part, these schemes are voluntary with incentives to participate. It is likely that they may become compulsory, with, and in some cases without, compensation for income lost. In such circumstances, conventional management practices no longer apply. The environmental constraints may be imposed in terms of the use of inputs and practices such as nitrogen use, or on limits on environmental impacts such as leachate in ground or surface water. In either case, farmers will have to record and report that they are meeting the set criteria. IT has a role to play for monitoring and confirming compliance.

Constraints on farming for environmental purposes initially are likely to come in technological packages, similar in concept to the blanket prescriptions of the conventional approach. Environmentally sensitive low input - low output systems are unlikely to sustain a viable farm sector without significant levels of income support. However, IT could help to formulate and implement locally relevant farming strategies, which, through better-targeted input use, achieve financially viable farming systems, which simultaneously comply with environmental quality criteria. In its most sophisticated form, IT would involve instrumentation and control for spatially variable field operations. Thus, IT by reducing the business risk inherent in environmentally sensitive farming systems, and providing the means of protecting environmental quality without compromising financial performance, could provide the key for sustainable agriculture in European arable farming.

Information Technology and Management

Managers are increasingly using information technology to help them improve their decision making capabilities, particularly when trying to manage the inherent variability found within the crop environment. One example is called Precision Farming, which is a systems approach to managing crops and land selectively, according to their needs. It utilises expertise from many disciplines and integrates the latest information technology tools and techniques to enable farm managers to get a better understanding and control of their fields.

Variabilities in Management

Management is the essential factor to achieve a stated outcome for the farm. Managers should identify their own strategies and practices that allow them to deal effectively with the variability found on their farm.

Three types of variability have been identified. The first type is spatial variability, which can be seen as changes across the field. An example would be where one side of the field yields higher than the other side.

Temporal variability is identified when parameters change over time. This can be seen when a crop starts by growing well but results in a poor yield.

Predictive variability, is the difference between what the manager predicted would happen and what actually happened. The classical example of predictive variability is where the manager predicts that a certain yield will be achieved if a certain amount of fertiliser is applied, but the crop does not achieve it because the weather changes. Each type of variability must be measured, assessed and possibly influenced, according to how significant it is.

Measuring variability: The first stage in the process is to measure important factors that indicate or affect the efficiency of the growing crop. The two main approaches are to create yield maps through instrumenting the harvesting system or assessing soil parameters by sampling. Both techniques give information about different parts of the cropping system. Yield maps are historic and cannot be used while the crop is growing, but record the actual yield during harvest. Soil sampling can be expensive but many soil parameters such as texture and horizon depths do not change over time, so are a good investment.

Measuring soil nutrient status must be treated with care as repeatability, let alone accuracy, is difficult to achieve. Sampling strategies based on a simple grid tend to be expensive and better-targeted sampling techniques are being developed. Asset surveys can also be carried out to record physical features, such as field and crop boundaries, high trees that may cause shading, compaction in gateways, etc. Other high-density measurement techniques are rapidly becoming more important such as Remote Sensing and ADP (Aerial Digital Photography) or non-contact sensing (e.g. Electro-magnetic induction). ADP can give real-time information of the crop canopy and allow management to be modified while the crop is growing.

Significance of variability

The spatial variability must be assessed for significance to the manager. Normally this is done by looking at the spread of the data or seeing if the extreme values lie outside acceptable thresholds, such as indices for soil nutrients. In Britain, one technique is to reclassify yield data into 'gross margin' maps (Fig. 1).

This technique deducts the variable costs from the income, which varies spatially with the yield, resulting in a gross margin map that shows which areas generated more income than others did. Some gross margin maps have shown areas that actually loose money. The temporal effect of weather and prices are shown in Fig. 2.

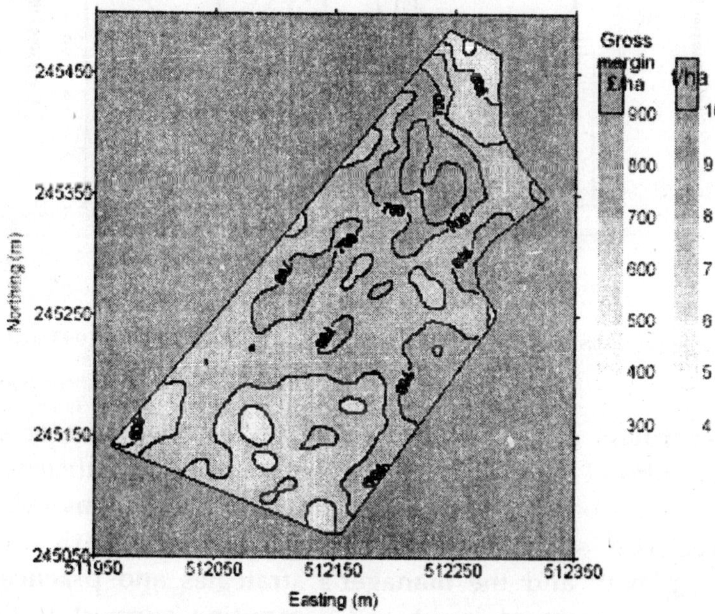

Figure . 1. Gross margin map with yield scale

The bars in Fig. 2 show the change in average gross margin over 6 years. Superimposed are the whiskers denoting maximum and minimum gross margins experienced in that year. The line denotes the average yield for that year and the selling price is included in brackets. Most traditional systems over-apply inputs such as seed, spray and fertiliser to reduce the risk of crop failure. With better assessment techniques, the inputs can be reduced or redistributed to optimal levels and the risk of failure can be managed. This will result in making the system more efficient.

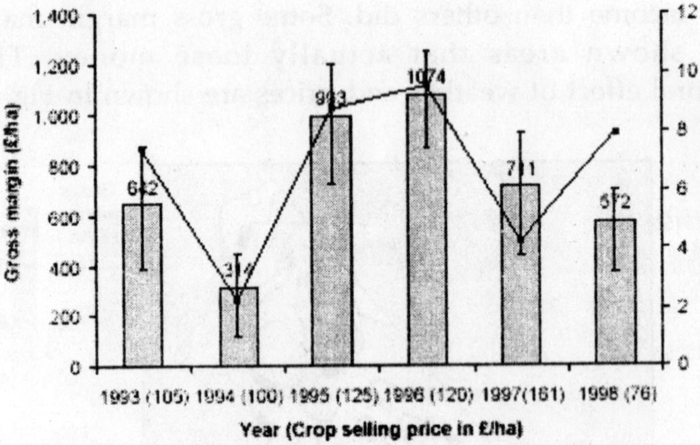

Figure 2. Gross margins and the in-field variability between 1993 and 1998

Regardless of the country and crop, management of an agricultural system is complex. To improve the efficiency, computer based Management Information Systems (MIS) must be sophisticated enough to deal with this complexity and the manager's strategies and practices. The management input and computing support is the same in each country and each crop. Some crops may well have special considerations that should be taken into account when designing the MIS, such as planning the harvesting logistics when supplying crop to a processing factory.

The size of the management unit in the treatment map depends on the ability to measure and manage it. Although a DGPS may be accurate to within one metre, a combine header may be ten meters wide, and the spreader width may be 24 meters. The management unit will be limited by the spreader width. A draft methodology for dealing with in-field variability has been proposed.

The ultimate goal is a full understanding of the whole process and rational decisions being made following the management philosophy. This will never happen, as there are too many unknowns in a very complex process, the weather being the most obvious. Nevertheless IT can still automate many of the repetitive tasks and help to provide accurate information at the right time. Levels of automation are bound to rise, as are the use of computers and controllers. The trend is towards a better understanding and a smaller unit of treatment.

Information sources will be on-farm as well as offfarm. Examples of on-farm sources are; yield mapping, weather station, quantities held in storage bins, weigh bridge, etc. Examples of off farm sources are; market prices, pest prediction models, expert advice, etc. Both of these sources could be integrated into a system that allows the manager easy and timely access precisely when the decision must be made.

9

Foliar Fertilisation

Foliar fertilisation (or foliar feeding) entails the application-via spraying-of nutrients to plant leaves and stems and their absorption at those sites. Used in both conventional and alternative production systems, it is a viable (though somewhat controversial) means of enhancing crop nutrition.

Foliar feeding has been used as a means of supplying supplemental doses of minor and major nutrients, plant hormones, stimulants, and other beneficial substances. Observed effects of foliar fertilisation have included yield increases, resistance to diseases and insect pests, improved drought tolerance, and enhanced crop quality. Plant response is dependent on species, fertiliser form, concentration, and frequency of application, as well as the stage of plant growth.

Foliar applications are often timed to coincide with specific vegetative or fruiting stages of growth, and the fertiliser formula is adjusted accordingly. Applications may also be used to aid plants in recovery from transplant shock, hail damage, or the results of other weather extremes.

In terms of nutrient absorption, foliar fertilisation can be from 8 to 20 times as efficient as ground application.

However, this efficiency is not always achieved in actual practice. Often, failures result from inattention to the principles of foliar feeding. Other causes of failure include application of the wrong spray mix, or of the right mix at the wrong time. Judging what foliar materials to apply and at what plant stage to spray them appears to be as much art as science.

Because of the variability in research results and practical field experience with foliar feeding, opinions on its usefulness vary in both conventional and alternative agriculture circles. There is general consensus, however, that foliar fertilisation should not be considered a substitute for a sound soil-fertility program. For operations seeking to farm more sustainably, this includes some combination of compost, livestock manure, green manure, cover crops, soil-applied rock minerals, and well-planned crop rotations that include legumes.

One of the touted benefits of foliar fertilisation is the increased uptake of nutrients from the soil. This notion is based on the belief that foliar fertilisation causes the plant to pump more sugars and other exudates from its roots into the rhizosphere. Beneficial microbial populations in the root zone are stimulated by the increased availability of these exudates. In turn, this enhanced biological activity increases the availability of nutrients, disease-suppressive biochemicals, vitamins, and other factors beneficial to the plant. It is this rationale, in good part, that reinforces the use of foliar fertilisation in organic agriculture, where the philosophy of feed the soil, not the plant prevails.

While foliar fertilisation is being used on a wide variety of crops, its economic value is generally deemed greater for horticultural than for agronomic crops. This is because horticultural crops are of higher value and their nutrient status is more carefully monitored. At present, for example, foliar sprays are commonly recommended to correct zinc deficiencies in grapes, to control bitter pit and cork spot in apples, and for general supplementary nutrition in strawberries.

Overall, the economics of foliar fertilisation is dependent, first, on how successful applications are and, second, on whether or not the same nutrition might have been supplied more economically through another means. Because weather can be a factor and because circumstances differ widely among farms and farm managers, there is no simple determination. The individual grower must decide based on need, and monitor for indicators of success.

Foliar Fertilisation and Pest Resistance

It is a basic tenet of organic agriculture that properly nourished crops will exhibit a natural resistance to insect pests and disease organisms. While traditionally most practitioners try to achieve optimum nutrition through direct soil management, many consider foliar fertilisation the final key to making some form of "induced resistance" a practical reality. There are several schools of thought regarding pest resistance and crop nutrition. Some suggest that well-nourished plants certainly have a better ability to tolerate pests and disease, but that there is no induced resistance; others deny any relationship at all between crop nutrition and the incidence of pests.

Some connections between pest problems and nutrient imbalances are documented. The presence of excessive soluble nitrogen in soils, for example, increases the nitrate and water content of plant cells. This is especially attractive to aphids, which thrive on plant sap.

Among the more popular theories is that plant sap with a higher level of soluble solids-measured as °Brix-is unsuitable food for plant-eating insects. Research on this theory is limited and far from conclusive. One California study, for example, failed to find a relationship between the °Brix of sap in commercial grape plantings and the presence of leafhoppers.

Some proponents of foliar fertilisation consider it an especially effective means of stimulating the natural defense mechanisms of plants. Studies to date are rather limited but have shown some positive results. Israeli research on corn using foliar sprays of phosphate and of trace nutrients demonstrated induced resistance to several diseases.

Basics of Foliar Feeding

For foliar fertilisation to work effectively, certain guidelines must be followed:

— To be efficient and to avoid crop damage, very dilute solutions of nutrient formulations are suggested. Sometimes as little as one cup to two quarts per acre of an active ingredient is all that is required to obtain the desired response. Highly concentrated sprays, especially those bearing inorganic salt-based fertilisers, have the potential to "burn" plant foliage. This is especially true of chloride salts.

— Spray-solution pH should remain in the near-neutral range. If pH adjustment is required, vinegar can be used to increase acidity, and baking soda to decrease it.

— In addition to pH, some other qualities of spray-water should be considered:

— *Cleanliness.* Small, undissolved particles can quickly clog nozzles-even those well-equipped with strainers can be plugged if enough suspended matter is present.

— *Chemical and disease contaminants.* Some water sources are contaminated and should not be used at all for foliar feeding. If there is concern specifically about disease organisms, the water can be treated effectively with a small amount of hydrogen peroxide.

— *Chlorine.* Chlorination of water removes harmful bacteria, but it also can kill beneficial organisms, which may be included in some foliar sprays. Allowing water to stand in an open tank overnight generally renders chlorinated water harmless to beneficial microbial mixtures.

— Best effect is achieved when foliar sprays are finely atomised. This can be managed by increasing sprayer pressure or by using a mist blower. Some advantage can be gained on boom-type sprayers by tilting the nozzles back to a 45° angle to allow the spray to drift onto the plants.

— Spray when wind is minimal. This is especially important with finely atomised sprays because they drift readily.

— Absorption is increased when sprays also reach and coat the undersides of leaves. This is where most of the plant's stomates are located.

— Always delay foliar fertilisation until air temperatures drop to 80°F or below. Absorption at higher temperatures is very poor because plant stomates are closed. Some of the most effective foliar feeding is done late in the evening or in the early dawn, when temperatures are right and wind is minimal.

— Absorption is further enhanced when weather conditions are humid and moist. The presence of heavy dew on the leaves facilitates foliar feeding.

— Addition of a surfactant to the solution decreases surface tension on the leaf and may increase absorption.

— Take note of possible chemical interactions among foliar fertilisers. Some materials are incompatible and should not be mixed together. They may create precipitates that tie up the nutrients and clog nozzles. Many product labels warn of incompatibilities. If there is no information, mix relative quantities of the materials and water in a jar and shake it. If there is no precipitate, there should be no problem.

— For convenience and cost savings, foliar fertilisation can sometimes be combined with a pesticide application. However, timing conflicts and material incompatibilities can make combining sprays unwise. Be certain to read all product labels and do the jar test if uncertain.

Two technologies that appear especially applicable to foliar fertilisation deserve to be mentioned. The first is the use of electrostatic sprayers, which impart a charge to the spray particles and cause them to adhere more readily to plants. The second technology, known as Sonic Bloom, uses sound to increase the leaves' absorption of nutrients.

Formulating Foliar Sprays

Foliar fertilisation is often attempted without clear objectives. Sometimes the grower has been convinced-perhaps by a salesperson-that spraying a particular product will be good for the crop. Sometimes it is simply a matter of faith.

Foliar fertilisation need not be undertaken in such a haphasard manner. There are several methods for determining need, possible benefits, and what materials to apply. These methods range from conventional analytical approaches to some that are best described as metaphysical. The most common farmer-friendly approaches are discussed here.

Deficiency Symptoms

In areas where crop production has continued for some time and where the interactions of particular crops and soils have been well studied, certain nutrient deficiencies are predictable. Where these deficiencies involve secondary nutrients and micronutrients, foliar feeding often becomes the preferred means of correction. For example, foliar feeding is routinely used in some regions to manage zinc deficiencies on pecan crops. Likewise, calcium sprays have often been recommended as one means to prevent blossom-end rot in tomatoes. The decision to spray in such cases is basically the result of past experience, often bolstered with soil test information and/or observation of symptoms in the field.

Plant Tissue Tests

Tissue nutrient tests are much more reliable than past history, plant symptoms, or soil tests for establishing whether nutrient deficiencies exist. They can, however, work hand-in-hand with any or all of these. In most high-value horticultural crops on large acreage (commercial pecans and tomatoes included), tissue testing is routine. This is perhaps the most commonly accepted means of identifying nutrient deficiencies. The suitability of foliar feeding as a corrective agent usually depends on the deficient nutrient. For example, it is more common to use foliar feeding if a micronutrient is needed than if nitrogen, phosphate, or potash are deficient.

Many soil testing services also offer tissue testing. The same is true of many state-run laboratories. The University of Arkansas, for example, has a program specialising in the foliar analysis of blueberries-a crop of significant value in the state. Be aware that accurate foliar analysis depends on good sampling and handling procedures. While instructions should be obtained from each laboratory or consulting service in advance, some good general advice is provided in the enclosure entitled Leaf/Petiole (Stem) Sampling.

Refractometers

An instrument that some growers find useful in combination with foliar feeding is the refractometer. Refractometers are low-cost hand-held tools that measure the dissolved solids (mostly sugars) in plant sap, by observing the bending (refraction) of light as it passes through the liquid. The higher the percentage of dissolved solids present, the better nourished the plant. As indicated earlier, many proponents of foliar fertilisation associate higher °Brix readings with pest resistance.

The process for using the refractometer to guide foliar feeding is relatively simple. Begin by extracting a sample of sap from one or more crop plants. This is accomplished by squeesing the leaves or stems with a garlic press or some other tool. Place the extracted sap in the refractometer to measure and record the reading. Then, use a spray bottle to mist a small sample of a foliar spray blend onto the plants. After a short time, extract and measure another sap sample. If the °Brix has increased, foliar feeding with that specific blend is advisable. By testing several possible blends using several spray bottles, you can determine the best spray to use. Of course, a control spray of water should be tested at the same time to discount any changes not caused by the fertiliser.

Radionic/Radiesthetic Analysis

Radionics and radiesthesia (dowsing) are metaphysical approaches to determining the need for foliar fertilisation and formulating the fertiliser sprays to be used. Despite the pseudoscientific nature of these modalities, they apparently are used with significant success by a surprising number of farmers. This writer had exceptional results using radionics to guide foliar feeding of commercial blueberries and blackberries in the late 1980s.

Fertilisers for Foliar Fertilisation

Synthetic Fertilisers

Most soluble conventional fertiliser materials can be used for foliar fertilisation. Hot mix liquid and dry soluble formulations are usually preferred, as they are designed to be dissolved in water and contain few contaminants. Fertilisers containing significant amounts of chlorine, however, should be avoided to reduce the chances of plant damage. Note that synthetic fertiliser materials are not permitted in organic production.

Organic Fertilisers

Fish-based fertilisers (fish emulsion or fish powder) and seaweed (soluble seaweed powder or seaweed extract) are among the most common foliar feeds in organic farming, applied either separately or in combination. Information on these fertilisers is available, as both have been in general use for many decades. Compost tea has become popular as a foliar spray material because of its nutrient content and disease-suppressive characteristics.

Other soluble organic materials and those from which extracts are easily made include spray-dried blood, bat guano, worm castings, manure teas, humates, molasses, milk, B vitamins, and herbal extracts of plants like stinging nettle and horsetail. Several enclosures are provided that give further detail on these and other organic foliar fertiliser components.

Organic growers should be cautious when purchasing commercial foliar feeding products. Not all are cleared for certified production. Some have been blended with conventional fertiliser materials. Certain stimulants, biocatalysts, and other materials are also prohibited, often because they contain or are derived from genetically engineered organisms.

Crop Manipulation

Fertilisation strategies can influence flowering, fruit set, fruit size, the amount of vegetative growth, and other plant characteristics. By carefully choosing the components of a foliar or sidedress fertiliser, the grower can "nudge" a crop toward earlier, heavier fruit set, or discourage fruiting-an advantage when producing greens or a forage crop. This concept is fairly well recognised in the conventional agricultural community. Many citrus growers, for example, are known to foliar feed with fertiliser blends dominated by potassium and nitrate-vegetative-growth-enhancing nutrients-to increase fruit size after the crop is well set. Generally speaking, fertiliser blends dominated by potassium, nitrate nitrogen, calcium, and chlorine tend to promote vegetative growth and fruit size. Blends dominated by ammonium nitrogen, phosphorus, sulfur, and manganese encourage the setting of fruit and seed.

While this knowledge gives the farmer more management options, one should not assume too much when trying to manipulate crop performance. All crops-but especially fruit crops-will need certain amounts of both growth- and fruit-enhancing nutrients throughout the season. Imprecise timing, or attempts to tip the balance too far, can yield disappointing and costly results. Note, too, that these manipulations can be especially difficult for certified organic producers to attempt with the smaller array of soluble fertiliser materials available for their use.

Hydroponic Foliar Fertilisation

The most commonly used method of hydroponic plant fertilisation is through a nutrient solution applied to the root zone of the crop. While plant root systems are in the most part efficient at absorbing mineral nutrients, certain conditions can prevent optimal uptake rates of some of the elements plants require. When plants are stressed for some reason, have suffered root death or damage, are showing a nutrient deficiency or are being established from cuttings, then foliar feeding becomes a particularly useful method of nutrient application. Foliar feeding, provides nutrients through the foliage of the plant which has the ability to absorb and translocate certain minerals within plant tissues and this is a technique which growers can use in many situations.

Benefits

In soilless systems such as hydroponics, many nutrient interactions can occur within the root zone which makes it difficult for the plant to absorb certain minerals. Even well run hydroponic systems can become deficient in nutrients—either because of nutrient depletion, antagonism between certain elements, or due to elements becoming 'bound' and therefore unavailable for plant uptake. Imbalances in the combination of nutrients, pH levels which are too high or low for maximum plant uptake and poor physical properties of the media surrounding the root zone, including oxygen starvation are more common in soilless systems than many growers realise. Furthermore, elements such as iron, an essential trace element, can not only be prone to becoming unavailable for plant use at moderate to high pH levels, but uptake by the plant can also be severely limited under certain environmental conditions such as cool temperatures. Iron chlorosis in many crops which are overly stressed by low temperatures can be a common problem where root uptake is the only source of iron for the plant.

Any situation which damages the root system or restricts its growth, development or physical process such as respiration affects the uptake of minerals. Plant pathogens such as fusarium pythium and phytophthora can not only rapidly destroy a crop, but low, less damaging levels can restrict function of the root zone to the point where mineral uptake is affected. While the crop may not show signs of severe infection, mineral and water uptake can be restricted and therefore crop yields and quality are affected. Other plant stress conditions such as anaerobic conditions in the root zone where oxygen is deficient, can limit nutrient uptake, with trace elements such as iron often affected to the greatest degree. Any other condition which stresses the plant—temperature stress, high or excessively low humidity levels, lack of light, high radiation levels, high plant densities, presence of pests or disease, will affect the efficiency of the root system in taking up mineral elements. These conditions are common and occur in many growing systems from time to time without the grower even realizing that plant growth and mineral uptake is being limited in some way. It is under these types of situations that foliar feeding has its greatest advantage. Since plant stress is dependent on a number of factors—many are environmental, which growers have limited control over, foliar fertilisation provides an 'insurance policy' against yield and quality loss from limitations in root mineral absorption and transportation.

Process of Hydroponic Foliar Fertilisation

Most leaves have stomata either only on the underside or on both sides of the leaf which enable gas to be exchanged for photosynthesis and respiration as well as releasing water vapour in stomatal transpiration. The leaf with its epidermis can also function as an organ that absorbs and excretes water and substances which may be dissolved in it.

Since foliar absorption is limited because of the relative barrier of the cuticle it is not possible to solely feed plants via the leaves. For this reason the most effective use of foliar fertilisation is as a rapid and effective method of supplying the micro nutrients. It can, however also be used to satisfy acute needs with lower concentrations of macronutrients and biuret-free urea is often used to supply nitrogen via the leaves.

Nutrient Application

One very important criterion of the effectiveness of nutrient sprays is the rate at which the foliar applied nutrients are absorbed by the leaves and translocated within the plant. The uptake of nutrients is further affected by a number of interacting factors of which only part are known at the current time—these are shown in Table 1.

Table 1 Influences determining the efficacy of foliar nutrient sprays

PLANT	ENVIRONMENT	SPRAY SOLUTION
Curricular wax	Temperature	Concentration
Epicuticular wax	Light	Application rate
Age of the leaf	Photo period	Application technique
Stomata	Air movement	Wetting agent
Guard cells	Humidity	pH
Leaf hairs	Drought	Polarity
Adaxial leaf side	Time of day	Hygroscopicity
Abaxial leaf side	Osmotic potential of root	Sticking ability
Leaf Turgor	Nutrient stress	Sugars
Surface moisture	Nutrient ratio	
Cation exchange capacity	Carriers, penetrates	
Nutritional status of the plant	Humectants	
Cultivar	Growth stage	

It would be difficult to ensure that all of these factors shown in Table 1, are optimal for foliar feeding at any one time, but some are more important than others. The use of a good quality, non ionic wetting or sticking agent is vital for foliar feeding. Wetting agents are necessary to ensure the adherence of droplets on difficult to wet leaves as well as assisting with the absorption of the fertiliser solution into the plant tissue.

The foliar fertiliser solution should then be applied as a fine mist until 'run off' so that the entire leaf surface is wetted. The time of day when the solution is applied is also important. Spraying early in the morning, while it is light, but temperatures are still cool, or in the evening is best and conditions that allow the leaf to dry rather than stay wet for an extended length of time is also important to consider. Foliar solutions should not be applied during hot, bright conditions, if the plants are wilting or under water/osmotic stress as the plants stomata are likely to be closed making application ineffective.

Foliar feeding can by carried out on a regular, weekly basis, or can be limited to the times when the crop comes under high nutrient demand such as early fruit set and heavy fruit loading. Often the greatest response to foliar feeding will occur during the active growth phases of plants. During these active growth stages, leaves show a particularly high efficacy for absorbing nutrients.

If a period during which the plants have difficulty in absorbing nutrients via the root system should coincide with a period when there is a particularly vigorous demand for nutrients, the result will be a significant loss in yield potential, without the grower seeing any visible signs of deficiency. Under such conditions foliar fertilisation can give particularly impressive results.

Role of Foliar Fertilisation in Hydroponic Crop

While hydroponic crops may appear to be supplied with optimal nutrition via a well balanced and formulated nutrient, they still benefit from the application of foliar fertilisers. Studies have shown that hydroponic crops such as capsicum, treated with a micro nutrient foliar applied solution, had an increase in fruit yield over control plants and also an increase in the compound capsaicin in the fruit tissue.

Hydroponically cultivated potato plants also showed similar results. Potato plants given foliar fertilisation treatments with a micro nutrient solution not only had a greater tuber harvest, but also higher dry matter of the whole plants. Rockwool grown tomato crops have shown both yield and fruit quality improvements when given a weekly foliar feed of a micro nutrient complex. These results would also be expected on a number of hydroponic crops with similar nutrient requirements and thus the process of foliar fertilistion is a cultivation technique that should be considered as more than just a 'quick fix' for mineral deficiency symptoms.

10

Impact of Climate on Agricultural Improvement

The whole foundation of present food and fiber production rests upon thousands of years of farmer trial and error through hundreds of human generations. Coupled with keen observations of climate/weather/crop/livestock interactions, farmer trial and error has produced most of our domesticated plants and animals, as well as many stable and productive agricultural systems. But the time-scale is daunting and the social cost incalculable, since for each incremental improvement there have been countless failures. Major objectives of agricultural research are to shorten the long time trajectory and reduce the social cost for the discovery and substitution of superior strategies and technologies.

Traditional crop and livestock technologies have evolved such that climate and weather information becomes an integral component of the total system. The often complex timing required to fit operations into the seasonal cycle and to avoid specific weather hazards is codified in ritual and custom. An excellent example is provided by Stanhill, who shows a close correspondence between the timing and sequence of special prayers for rain in the Jewish liturgical calendar, established some

200 years ago, and the present-day dependence of wheat yields in Israel on the critical early part of the growing season. Careful analysis of such ritual and custom can yield valuable information about climatic patterns and weather hazards in regions where instrumental data are lacking.

The primary objective of most traditional systems is to minimise year-to-year variation in productivity and, especially, to minimise the risk of total loss of crop and/ or livestock. But this stability exacts a price in terms of potential yields foregone. While the traditional-crop cultivars and livestock breeds are hardy and resilient to the impact of climatic variation and weather hazard, they are usually low-yielding and incapable of an economic response to added inputs or new management techniques. So well adapted and finely turned are some of these systems that sudden change in any major component cultural, biological or physical can cause massive disruption

Modern, high-technology agricultural systems usually aim at maximising the expectation of profit or yield, using crop cultivars and livestock breeds that have been produced to fit specific environments and markets, and inputs of fertilisers, herbicides, pesticides and machine energy. Normally, buffering reserves of food and/or capital permit such systems to adopt higher-risk strategies that lead to much greater production on a time-averaged basis. Acceptance of an occasional crop loss or uneconomic yield may be normal for a capital-intensive farmer, but it is unlikely to be acceptable to a subsistence farmer. These distinctions are important when globally assessing climatic impact on agricultural production. Estimation of risk is an important component of agricultural analysis, but the level of risk acceptance must be taken into account.

Evidence suggests that the earliest cultivators recognised that natural vegetation could provide an index of site quality for their crops. Modern man has added a little to this basic concept the transfer of information by analogy. The basic hypothesis is that all occurrences of a defined class of land will respond similarly to any imposed treatment. Climate, soil, vegetation and multi-attribute land classifications exemplify this approach. These classifications then provide a basis for the selection of 'representative' experimental sites and for the subsequent extrapolation of results to other sites that share similar properties. Most current agrobiological research rests firmly upon such a foundation. Impact assessment can also employ analogues as simulators of future climate change. Analogues are sought that have the characteristics of the simulated or predicted climate, and the crop types or natural vegetation found within such climatic locations serve as the simulated or predicted vegetation.

Such a strategy is logical when the understanding of functional relationships between site attributes and crop response is qualitative or lacking, since this understanding is not required for classification. However, the better such knowledge is, the better will be the choice of attributes and the resultant classification. The thematic maps produced form a major component of an information system that is relatively inexpensive and highly portable. Usually agricultural research strategy is based upon climate and/or soil classifications, while forestry and pastoral research may place additional emphasis on vegetation classification.

Major deficiencies of the analogue approach reside in the static, multi-attribute character of the classifications upon which it is based and the implicit assumptions of covariance of attributes within mapped units.

Compounding this is the location, season, cultivar and management-specific nature of the results from traditional agronomic experiments. For the individual farmer, the ideal location for an experiment is on his or her farm or as close as possible, since successful extrapolation of results is observed to hinge on proximity to the experimental site. Accordingly, there is continuing sociopolitical pressure to extend the network of research stations and experimental sites.

Given that agricultural research strategy will remain firmly based upon concepts of transfer of information by analogy for some time to come, there is ample scope for improvement and refinement of climate, land and soil classifications and for optimisation of experimental networks. Numerical taxonomic analysis and/or pattern analysis provide an array of classifications. The more relevant the attributes are, the better is the chance of a useful result. Commonly, the attributes measured represent a compromise between sampling density and site measurement.

Agroclimatic and bioclimatic classifications provide illustrations of the basic problems inherent in the analogue approach. While there is a considerable literature on bioclimatic classification and associated technologies, most of the general, multipurpose classifications are of limited use in solving specific problems. More direct coupling of climate data and specific crop response data offers prospects of more relevant classifications.

Agroclimatological zonation has always been particularly attractive to data-poor developing countries without an extensive network of meteorological and agricultural stations. They construct broad agroclimatological or agroecological zones using available temperature, precipitation, and soil and crop data. One example that illustrates this approach is the work of Jaetzold, who was asked by the Kenyan Ministry of Agriculture to provide a highly differentiated system of agroecological zones to help decide which variety of crops would best utilize the natural potential of a given district of arable land.

Analogue methods are currently in use to predict potential agricultural change under conditions of climatic change, particularly CO_2-induced change. At a continental scale, a characteristic bioclimatic classification is that of Holdridge, based on temperature, precipitation and evaporation. This classification currently is being used to analyse the impacts of a CO_2-induced climate change. At other scales of analyses in such data-rich countries as the United States, shifts in the boundaries of the corn belt have been projected for a hypothetical CO_2-induced climate change by Newman and Blasing and Solomon.

Regression Techniques

Since the pioneering work of Hooker and, more particularly, Fisher, both simple and multiple regression techniques have become standard practice in the agricultural and biological services, both for exploratory data analysis and for prediction. Such empirical/statistical modes of crop/weather analysis dominate the voluminous literature of agroclimatology.

Correlations have been established among a wide range of environmental variables and virtually all aspects of crop production. The utility of many such correlations under the conditions in which they were developed is not doubted, but the majority have limited application. Generally, the simpler the measure used, for example, total annual or seasonal rainfall, or annual mean air temperature, the more location-specific is the relationship.

Data on crop yield, temperature, and precipitation for a given district are the central components in statistical regression models. As reported by Waggoner, the model for a particular crop in a particular area is generally expressed as

$$Y_i = a + b_1 t_i + b_2 X_{2i} + ... + b_n X_{ni},$$

where

Y_i	=	estimated yield in the ith year;
a	=	intercept;
t_i	=	surrogate for technology in the ith year;
b_1	=	coefficient representing the effect of technology in quintals/hectare/year;
b_2 to b_n	=	coefficients representing the effect in quintals/hectare/ unit change in the weather;
X_{2i} to X_{ni}	=	weather variables such as precipitation, temperature, potential evapotranspiration (PET), and evapotranspiration (ET) in the ith year.

Thus, to quote Waggoner, 'by multiple linear regression the effect of weather, factor by factor, on the yields of crops is distilled from a history of weather and yields, accumulated for decades by faithful observers'.

Greatest success with this approach has been achieved where one, or perhaps two, environmental factors dominate crop performance. Thus, in high latitudes, temperatures that determine length of growing season may be most critical; in midlatitude, subhumid to semi-arid regions, rainfall amount may be most critical; and in the humid tropics, with high rainfall and near-continuous cloud cover, solar radiation receipts may be most critical. The use of simple variables, such as monthly mean or seasonal mean precipitation and temperature, is common in large-area studies where the availability of data is restricted and/or computational limitations occur. Examples of successful applications of such analyses using simple data are provided by Thompson in his analysis of weather and technology in the production of wheat, corn and soybeans in the United States.

Usually, greater precision and generality are achieved if the primary climatic and/or soil data are transformed into indices that are more closely coupled to system performance. The derivation of crop/water stress indices from water-balance models has proved successful in the explanation of large-area variation in wheat yields. Subsequent developments in the understanding of crop response to water stress have led to the formulation of empirical/statistical functions for a very wide range of crops.

Inherent problems of the empirical/statistical approaches based on regression-type analyses relate to explicit assumptions of linearity of response, the interdependence of so-called independent variables, implicit assumptions that correlation implies causation, and, inevitably, the location, season, cultivar and management-specific nature of resultant relationships. With care, these limitations can be minimised, and new techniques offer prospects of improved linear and nonlinear estimation procedures.

Regression techniques form part of the basic armory of methods used in the newer systems-based research strategies, both for exploratory data analysis and to quantify functional relationships between variables. Studies employing regression have been developed for a variety of questions about agriculture in relation to climate variability and change. As in the studies of Thompson cited above, a major objective has been the identification of systematic patterns of yield variation. For example, Newman and Waggoner have examined whether there has been an increase or decrease in crop yield variability between the 1930s and the present and whether there is a correlation in behavior between major grain-producing regions. In general, they find relative variability has decreased somewhat, while variability expressed as absolute yield has increased.

Regression-centered studies were extremely popular during the 1970s as nations and regions tried to assess their vulnerability to climatic variability. Good examples of national studies are found in McKay and Allsopp for Canada, National Research Council for the United States, and Takahashi and Yoshino for Japan and southeast Asia. For the World Climate Conference, Oguntoyinbo and Odingo explored relations among agriculture, climatic variability and various other causal factors in Africa, while Mattei, McQuigg, and Fukui prepared perspectives on the semi-arid regions, temperate regions and humid tropics, respectively.

A recent application of a regression approach to impacts of long-term climatic change is found in Waggoner. Waggoner explored what the impacts of an annual average warming of 1°C and decrease of 10 percent in rainfall might be on wheat, corn and soybean production in the US Middle West, under assumptions of no adjustment or adaptation. While such studies are obviously tentative, they may give a sense of the order of magnitude of prospective impacts. These empirical/statistical models, like those that relate crop yields to atmospheric pressure patterns may give sound results, providing that climate change does not induce atmospheric patterns that lie outside the range for which the yield regressions have been developed. This was recently illustrated by Santer, using the work of Hanus, whose models for winter wheat usually predict actual West German yields within 4 percent, except for the anomalous year 1976. In that year precipitation in the key month of June was a quarter to a half of the long-term average for most of West Germany, effectively outside the range from which the regressions were developed. Applying scenarios derived from two general circulation models of CO_2-induced change to the Hanus models evidenced similar problems, leading Santer to conclude that simple multiple linear regression appraisal may be basically unsuitable for making any credible assessment of the impacts of climatic change on crop yields.

Multivariate Techniques

These techniques represent a logical mathematical development of regression-based analysis in that they permit simultaneous analysis of sets of interacting variables; multivariate analysis, factor analysis, principal component analysis, and principal coordinate analysis are examples. While of principal value for data reduction and in exploratory data analysis, they can provide a basis for useful prediction and partial explanation. However, the derived principal component values, eigenvalues, or factor scores remain static representations of dynamic interactive processes, and their validity is strictly limited to the data domains used to develop them. While it is possible to determine the relative contributions of separate variables to the derived composite axes, it is often difficult to provide any physical explanation for their operation.

Sophisticated applications of multivariate approaches are found in the work of Monteith. Monteith developed a simple model of crop growth for tropical crops, including climatic factors, and has modified it for application in other climatic zones. For example, Monteith (employed the model to estimate the ultimate limit set by climate on the productivity of British farms, defining efficiency of crop production in thermodynamic terms as the ratio of energy output (carbohydrate) to energy input (solar radiation). Temperature and water supply were found to be the main climatic constraints on efficiency. Over most of Britain, where radiation and thermal climates are uniform, Monteith found rainfall to be the main discriminant of yield between regions.

Another example of climate impact studies employing multivariate analysis is found in the discussion of Gifford and Monteith, who have debated maximum growth rates for crops like maize, sorghum, millet and sugarcane (the so-called C4 crops), that grow well only in warm climates, and potatoes, sugar beets, rice, kale and cassava, growing either in warm or cool climates. Such studies come close to attempts at full simulation of plant behavior. The simulations, discussed below, are based more on hypothesised causal relations than on observed historical relationships, but the distinction is often blurred.

Systems Analysis

Systems analysis is concerned with the resolution of any complex system into a number of simpler components and the identification of important linkages between them. Subsequent synthesis in the form of a symbolic representation (diagrams, flow charts) that is formalised as a set of logical statements or algorithms and mathematical formulae leads to the construction of a model of the real system. This model is then tested, using appropriate data inputs, and revised and developed to the point where it successfully simulates the behavior of the real system. Beyond this point the model can be used to predict probable outcomes of experiments performed on the system. For many complex systems simulation-based experiments represent the only possible means of examining perturbations to the system. Climate and weather systems and biological systems are examples of complex systems that can benefit from the systems approach.

The process of formal analysis of any complex system yields other benefits in that it formalises what is known about the system, identifies the more important components and processes and significant constraints on performance, and provides a suitable vehicle for testing alternative management strategies and the 'what if' questions posed by scenarios of global climatic change. Despite the proliferation of systems terminology and its ready incorporation into program titles and statements of research objectives, few agrobiological research strategies are truly systems-based.

Essentially, this is because a systems-research strategy demands interdisciplinary teamwork that transgresses the boundaries of long-established discipline and subject-specialist groups and competes with them for scarce resources. As in any research team, leadership is critical, but special problems relate to the maintenance of subject-specialist skills of team members, peer recognition, and the allocation of rewards and responsibilities. Most successful systems-research groups have evolved around a nucleus of a small group of committed scientists rather than through the imposition of a systems-research structure by higher authority.

Agroclimatic Synthesis

Past, present and, predictably, future modes of agroclimatic analysis and synthesis are inextricably linked with the broader goals of agricultural, pastoral and forestry development and land and water resource management. What, then, are these goals? Obviously, these will vary from field to field, farm to farm, region to region, nation to nation and continent to continent. But an ecologically sound and socially responsible goal would be to develop technologies that lift production, while maintaining the long-term stability of the land and water resource base. Implementation of such technologies rests, ultimately, with the individual land manager, whether profit maximiser or subsistence farmer.

The central problem, then, is how to prescribe a technology that is relevant to the land, labor, capital and management resources of the individual land manager. If it were possible to predict the performance of any production system at any location given a specified minimum set of soil/crop/weather-management data, then it would become possible to prescribe appropriate technologies. This ultimate objective is attainable, but that it requires a shift in emphasis away from reductionist/analytical research towards holistic/synthetic research. Of course, both types of research are necessary; they are complementary rather than competitive.

While much of current agrobiological research remains committed to the statistical differentiation of treatment effects on specific crops and cultivars, and on specific soils at specific locations in specific seasons, there are encouraging signs that greater efforts are being directed toward solution of the general problem. Of necessity, the earliest attempts at modeling agricultural systems were concerned with specific processes and/or specific products within restricted environments. Now, wide ranges of fully operational models of crop, livestock and forest systems are available or are under active development. Some, at least, are capable of general application across a spectrum of crops and environments.

Active development of simulation modeling and access to sophisticated computer technology are not in themselves, however, sufficient evidence of adoption of a systems-research strategy that aims at prescriptive rather than descriptive solutions. If, as stated, the primary objective is to develop models capable of providing both general solutions and location-specific prescriptions, then the focus must be on the balanced development of two interactive components (a) the models, and (b) the data base. The models should be fully accessible and operational at all times, yet capable of continuing improvement in logical structure and function.

The use of subsystem modules facilitates flexibility. The data base contains only those physical, biological, social and economic data that are specified as necessary for implementation of the model. Balanced development of both components is critical. Without access to prescribed input and test data, models cannot be implemented. On the other hand, the acquisition of data for which no specific requirement exists can be wasteful of scarce resources.

Agricultural Systems Models

Models should be simple; most would agree that they should be testable and capable of improvement after testing. Following the last two decades of development of models of ever-increasing complexity, there are signs of real efforts to return to simpler and more general crop models. In many cases, when it comes to implementation of the model in the real world, this is no more than a response to the reality of data limitations. In other cases the ability to simplify and generalise functional relationships has been a consequence of real improvements in understanding through basic research. Increasingly, models are being constructed using modular concepts that permit rapid updating and transfer of modules between models. Thus, it is possible to recognise common modules for, say, the water balances processes in many different models.

Van Dyne and Abramsky reviewed agricultural systems models then extant in some detail. They defined agricultural systems broadly and included those renewable natural resource systems and artificial biological resource systems that are directly under the control of man. They tabulated characteristics (structure; mathematical techniques; time-step; number of driving variables, processes and parameters; programming language; availability of flow charts and source code; stated objective and actual use and comments) for 90 papers and referenced 66 others in modeling applications to dairying and beef feedlotting, grazing sheep, irrigation, pest control, crop growth and yield, forestry, fisheries, wildlife management and agricultural economies.

A general conclusion was that most models were poorly documented and thus not readily transferable. Many models provided a useful learning experience for the developers, but few become fully operational in day-to-day planning and management. It is hoped that this situation is changing, but Van Dyne and Abramsky's recommendations on documentation bear repeating:

1. specify objectives,
2. specify the hypotheses or assumptions used,
3. specify the general mathematical form of the model,
4. list the specific driving variables,
5. list the state variables used,
6. show a diagram of the model structure,
7. list equations or functional relationships,

8. list the computer code,
9. provide adequate comment statements,
10. report model deficiencies and limitations,
11. provide test samples of data used, and
12. provide output of a test case.

Although new scientific journals are catering to the specialised requirements of agricultural systems models, few can afford to publish lengthy computer program listings.

Various attempts have been made to classify the wide variety of models in use and under development, ranging from basic purpose to type of formulation. Van Dyne and Abramsky in their comprehensive review found that most agricultural system models were deterministic; the same would be true today. Stochasticity is introduced through the use of historical, synthesis or real-time weather and other environmental data. Ideally, the output should be expressed as a probability function.

Agricultural production systems of all kinds and at all scales draw upon three basic types of resources. Natural resources include the physical and the biological. Capital resources include items of plant and machinery, draught animals, fertilisers, herbicides, pesticides and so on. Human resources, the vital component, include all aspects of labor and management.

Many models have concentrated on the biological and physical processes that determine crop growth, development and yield and/or animal production, and others have focused on optimisation procedures that explore economic and/or social objectives. Few, as yet, have attempted to encompass the whole spectrum of interactions and relational processes.

An interesting systems perspective, though not formalised in a mathematical model, is that of Anderson. Anderson, employing a general framework encompassing farm, regional, industrial, sectoral and national levels, has examined impacts of climate variations on Australian agriculture. Anderson begins with behavioral assumptions at the farm level, assuming first that satisfaction in farming depends on both personal characteristics and the probability distribution of financial performance. Financial performance depends in turn on allocative decisions, institutions and governmental interventions, prices and yields. The last two are, of course, in turn dependent on climate and a variety of other factors.

As Anderson reports, the effects of extreme climatic regimes have been sketched as unambiguously severe in affected regions. At high levels of geographic aggregation, impacts are often less severe because of the 'cancelling out' of good and bad experiences. McIntyre estimated that gross rural output in Australia in the drought years 1965 66 and 1967 68 fell by 11 percent and 15 percent, respectively, below smoothed trend values. There is debate over the multiplier effects of drought on the national economy, some results suggesting amplification and others dampening.

The Erosion-Productivity Impact Calculator (EPIC) model referred to above is one of the most comprehensive agricultural system models now operational, and it is undergoing continued development. It addresses a vital socioeconomic and biophysical issue soil erosion at national levels, but it could have addressed climate change and variation equally well. EPIC has been developed by a large modeling team formed within the Agricultural Research Service of the United States Department of Agriculture. The brief given was to determine the relationship between erosion and productivity for the whole range of soil and crop interactions in the United States. Williams et al. provide a detailed description of the model, which has physically based components for simulating erosion and plant growth and economic components for assessing the cost of erosion, determining optimal management strategies, and so on. The physical components include weather simulation, hydrology, erosion/sedimentation, nutrient cycling, plant growth, tillage and soil temperature.

For every model it is necessary to identify the minimum data set necessary for its successful development, validation and subsequent implementation. The early stages of conceptualisation, flow-charting, and programming usually rely on the current understanding of process, function and interaction as established by literature review. Any testing and validation relies on readily available data. Many modeling exercises never progress beyond this stage. Those that do soon realise that data limitations become the major constraint on further progress. This is particularly so when further development and testing require balanced and matching site/crop/weather-management data from a range of contrasting environments.

Ideally, specific field, laboratory and/or controlled environment experiments would be designed to acquire the necessary minimum data set in the shortest possible time with the least expenditure. In reality, most model building and testing in this phase have had to rely on a fortuitous assembly of data from a range of sources. Few existing experiments, unfortunately, can satisfy even the barest minimum requirement for balanced and matching sets of soil/crop/weather-management data.

Proposals that standardised data sets be collected from experiments are not new. A number of national and international research programs have adopted this principle. What is new about the concept of minimum data sets is that it arises directly from the adoption of a systems-research strategy. Each model specifies the minimum data set necessary for its development, valid action and implementation.

Review of existing operational models of agricultural production systems in general, and of crop systems in particular, indicates that many data items are common to all minimum data sets. In fact, a remarkable convergence in specification is under way. All share the need for primary weather data; terrain and soil attributes that modulate water balance and nutrient cycling; plant growth attributes that specify growth and development responses and yield accumulation processes; and management inputs that modify any of the foregoing. While it is possible to erect a hierarchy of minimum data sets with increasing range, precision, accuracy and frequency of measurement, at each level emphasis remains on a balanced monitoring of the whole crop system.

A new international program, International Benchmark Site Network for Agrotechnology Transfer (IBSNAT), was initiated by the US Agency for International Development in 1983 with the objective of developing a systems-based strategy for agrotechnology transfer. Through collaborative experiments in a large number of tropical and subtropical countries, standardised minimum data sets are to be collected for major food and cash crops.

Development of operational models of these crops will proceed in parallel. The field experimenter will have access to models that will run on his or her own data, and the modelers will have access to standard minimum data sets from a wide range of contrasting environments.

Useful and profitable interactions between the two groups can be expected to follow. Provided that the specific minimum data sets are monitored, no particular restrictions on experimental design need apply.

The IBSNAT program should help to remove constraints that currently hamper the further development and testing of operational models of crop systems. But what of the application of these developed models to real-world problems on an extensive scale? Such implementation requires that the appropriate model be coupled to a specified data base. Usually this will involve a climatic data file and either standardised or actual terrain, soil, crop and management variables. Obviously, the full flowering of these new technologies will depend on the coupling of an appropriate model to a resources data base that provides the necessary climate, terrain, soil and management data at the required scale and level of resolution.

New technologies currently available and under active development offer prospects of economical and efficient computer storage of geocoded data. Automatic digitising equipment can process and store contour maps as arrays of grid points. Sophisticated surface fitting algorithms can be used to reconstruct the contoured surface, but more importantly, to estimate slope, aspect, elevation, position in landscape, water concentration, and overland flow trajectories and other important terrain-related attributes. Combined with similar algorithms that permit estimation of long-term mean climatic data at any point, the stage is set for the derivation of necessary input data for models at levels of resolution that are useful.

The generation of stochastic sets of daily weather data based on statistical analysis of historical weather sequences and correlations between individual weather elements has been shown to be practicable. Initial attention focused on daily rainfall, but this has been extended to include other important elements such as maximum and minimum temperature and daily solar radiation.

Interpolation procedures can be used to generate satisfactory sequences of daily weather data at any geocoded point. The implications of this new technology for systems-based research and use of models in agrotechnology transfer are profound. Technically, there are no reasons why these techniques should not be extended to generate stochastic sequences of daily weather data for much of the surface of the planet. On land, the value of such a data base for land evaluation, planning and management would be incalculable. Perturbation of the basic equations could be used to generate notional climates in some postulated climatic change and coupled to crop models to provide 'what if ' answers to speculative questions.

11

Crop Marketing Services

Inputs, equipment, and crops are mainly private goods because most are excludable and rivalrous. Indeed, production and distribution of agricultural inputs and equipment, as well as crop marketing, are typical private sector businesses in most countries. In principle, it should be possible to decentralise these services in developing countries through market liberalisation, encouragement of private and voluntary sector organisations to enter service provision and improved efficiency and/or divestment of public sector organisations. However, for effective government disengagement, markets have to be well developed, there has to be an absence of general or specific market failures, and no other grounds for government intervention such as merit good status or income redistribution.

Factors Affecting Crop Marketing

Fertilisers and Chemicals

Fertiliser distribution is an ideal service for private sector provision, as fertiliser can be considered as a pure private good being both rivalrous and excludable. The demand for fertiliser and chemicals is a derived demand and will be influenced by the price and profitability of the crop as well as the price of the input. However, fertiliser use often increases the risk faced by farmers if rainfall is low and/or crop prices are below expectations. This will tend to make the demand for fertiliser unstable as well as seasonal. With imperfect and incomplete finance markets, credit-constrained farmers will find it difficult to purchase fertiliser unless new channels of credit provision are made available to them. This is one conceivable disadvantage of scrapping single channel marketing systems. Although a temporary problem, exchange rate devaluations in the early days of liberalisation combined with the removal of subsidies often led to substantial increases in domestic fertiliser prices and marked falls in effective demand.

For low-income countries it is sometimes argued that food security stimulated by fertiliser use on food crops generates positive externalities and thus, in principle, a case may be made for government intervention in the form of fertiliser subsidies. This is particularly true if the demand for fertiliser is price elastic and reducing the price of fertiliser leads to a marked increase in food production and food security.

There are two types of imperfect information relating to fertiliser and chemical use. The first relates to the quality of the products. The effects of fertiliser applications, in particular, are often not seen for several months. It is therefore quite feasible for an unscrupulous trader to sell inferior quality products to farmers. The second relates to appropriate advice relating to applications. If this is not available, input use is likely to be sub-optimal or even ineffective. However, for fertiliser applications this advice tends to be crop- and location-specific as it is closely linked to soil and climatic conditions. Like all information, it has public good characteristics that suggest that the government should be involved in its provision. With pesticides and herbicides, incorrect use can generate serious negative externalities in the form of threats to human health and environmental damage that justify some form of government intervention in provision.

The specific market failures relating to fertilisers and chemicals include:

— High working capital requirements for importation, wholesaling and stockholding due to seasonal demand
— Uncertain demand affected by anticipated weather and crop price conditions
— Low effective demand if farmers are cash- or credit-constrained
— Playing field not level for private sector if donors/government supply subsidised fertiliser
— Imperfect information relating to product quality and appropriate recommendations for use
— Possibility of negative externalities, particularly with pesticides and herbicides.

Planting Materials

Seed supply is a complex operation consisting of three major components:

— Research into new and improved types of seeds
— The physical supply of seeds
— Regulation including:
  — the management of variety testing in public sector plant breeding
  — variety regulation (registration, performance testing and release)

— seed quality control consisting of seed certification (verification of genetic quality) and seed testing (analytical purity and germination capacity).

In pre-reform situations public sector organisations are frequently responsible for all three functions, but there is no economic reason why this should be the case. Nevertheless, the nature of seed technology and the characteristics of the product suggest that the public sector is likely to continue playing a role in the provision of the research and regulation components.

Seed technology is rivalrous but only in the case of hybrid varieties is excludable. Open-pollinated varieties (OPVs) can be reproduced by farmers themselves and passed on through informal seed distribution mechanisms. This makes them common pool goods. While private sector firms are willing to conduct research into developing new types of hybrid seed, they will often not find it attractive to develop new OPV seeds. Exceptionally, they may have a clear advantage in terms of quality and reliability of supply that enables them to capture a significant share of the market.

In some developed countries plant breeders' rights give sufficient protection to make private sector provision of OPVs a viable proposition but it is debatable whether they would be appropriate or enforceable in developing countries. In small and poor countries with small potential seeds markets it is likely that the public sector, or non-profit organisations, will always play an important role to overcome this particular problem of market failure.

One problem facing potential entrants to the seed supply sector is the strong seasonality of demand that means large stocks of perishable seed need to be carried.

This ties up working capital but investment in adequate storage facilities is also required. The private sector also requires a level playing field and this is unlikely to be achieved if either the public sector or NGOs are subsidising seed production or sales.

Several aspects of seed regulation require at least some form of input from the public sector to overcome potential market failure problems. Seed certification must be sponsored by the public sector although it can be performed by a private organisation. The government must also regulate seed testing, although companies that are trying to establish a reputation for themselves are inclined to apply their own adequate seed testing procedures, and associations of seed producers may provide this service for their members.

Equipment Supply

There is a potential problem of market failure relating to imperfect and asymmetric information. As with fertiliser, it may be difficult for a farmer to assess the inherent quality of machinery and this creates particular difficulty when items involve substantial investment and purchases are only made infrequently.

There are three potential market failure problems with agricultural produce marketing that need to be taken into account when considering decentralisation. First, it is inherently risky because of unpredictable variations in output that cause marked price and income fluctuations. These in turn may cause food insecurity. Second, there may be positive externalities related to food security that justify some form of government intervention. Finally, spatial dispersion means that small and remote farmers may suffer from a lack of information and exploitation by traders.

There are a variety of market failure problems that affect the demand and/or supply of these agricultural support services. The most common concern situations in which one or another of the following features, or combinations of such features, prevail:

— Natural, de facto, or legal monopolies (present at any level of the transactions in the market for a specific product)

— Interlocked markets, when input suppliers/crop buyers extend credit to farmers in small fragmented markets

— Asymmetric information, when knowledge about price and quality of a product is not equally shared by traders and clients

— High transaction costs, due to poor transport and communications infrastructure, and an inability to obtain credit from formal institutions

— Inadequate rural financial services (farmers have no access to credit or other banking services)

— Inadequate or inaccessible market infrastructure which may mitigate risks and uncertainties (storage, communications, information, commodity exchange)

— Very high start-up costs for competitors (which make multiple entry difficult, inhibiting competition)

— Uncertain demand for inputs due to unpredictable climatic variability.

Decentralisation

In principle it should be possible to decentralise most input and equipment supply and crop marketing services in developing countries through market liberalisation, encouragement of private and voluntary sector organisations to enter service provision and improved efficiency and/or divestment of public sector organisations. This, indeed, has already occurred in many countries.

State 'shrinkage' has played a role in the measures undertaken so far and has included actions such as:

— Liquidation of autonomous crop marketing public agents

— Withdrawal of the extension services from input supply

— Liquidation of bankrupt agricultural banks heavily engaged in extending credit for input purchase to non-repaying farmers

— Halting subsidy programmes, which led to the liquidation of state supported cooperatives.

Even though the private sector has assumed a much larger role in input supply and crop marketing, governments at the central and local levels have an important role to play in creating an enabling environment for private sector activities.

Encouragement of Private and Voluntary Sectors

The development of a competitive market for agricultural inputs requires inter alia that:

— Equal opportunities to trade are given to all potential suppliers

— Adequate credit facilities are available to private traders

— Producers/suppliers selling products that embody special technical characteristics are protected against unfair competition.

— Farmers have an adequate knowledge about the quality of the products that they buy.

Naturally, for an effectively competitive inputs market to achieve an efficient allocation of resources in the agricultural economy of a country, an equally effectively competitive market for crops must be in place. Four main conditions must be met for this to happen:

— Deregulation and liberalisation of the national crops market, including imports and exchange rate and a least distorting and transparent taxation/ subsidy regime
— Adequate market information available to producers and traders
— Adequate access to credit by traders and producers, to enable the latter to finance their stocks on the farm prior to sale. This will improve their negotiating power vis-à-vis traders
— An adequate network of transport infrastructure in the rural areas.

Nature of Market Development

The development of functioning private markets is not a short-term objective. In many developing countries, particularly small countries at the lower end of the scale in terms of income, there is hardly what can be called 'a private sector' capable of important new initiatives. It takes time for this to come about. In larger countries with more developed economies, the private sector is often dominated by large traders who concentrate on import/ export operations and in supplying urban areas, where the volume of business is large and profit margins significantly higher than in rural areas. As a result, in many rural areas of larger and more developed countries the situation is not dissimilar to that of smaller and less developed countries.

Under these circumstances the policies aimed at developing functioning rural markets must be prompted by a long-term vision. New rules and other measures must be introduced and implemented in a coherent manner and on a permanent basis. This is a major challenge that is not always faced adequately. In fact, the persistence of the negative effects of market failures and the many factors related to the power configuration of the public agencies involved have often provided justification for continued government intervention in the agricultural markets as 'interim measures'. Some interim measures have actually represented partial reversal of reform policies, discouraged private initiatives and seriously jeopardised the effectiveness of the reform.

Private enterprise will respond over time to meet an effective demand for agricultural inputs. This will be generated from the development and promotion of crop technologies that bring about significant, not marginal, returns to farmers. This will often require actions by both the private and public sectors. Large trading companies engaged in export crop processing and input producers both have a general interest in promoting the growth of demand for inputs, because it brings about more crops to buy, not just more inputs to sell. These private enterprises may go beyond the assessment of existing markets and often have the resources to do so.

A dynamic fertiliser market, for example, needs to be fostered by continued research on soil fertility, on crop responses under different ecological conditions, on adequate crop protection practices and products, on variety development, cropping practices, etc. In the case of 'cash crops' input suppliers and crop buyers do get involved in research and extension. This is not only the case of large-scale multinational companies. In Kenya, for example, several medium-size private companies engaged in seed production and in export of horticultural products are now involved in research and extension activities.

To promote increased effective demand by commercially oriented farmers, a crop approach to research and extension is often sufficient to identify which technical packages interest producers, stimulate more demand for inputs and production of more crops of interest to buyers. For the other category of farmers, matters are much more complex. Here, a farming system approach, continuous participatory diagnosis of problems and on farm adaptive research are required.

Support to Alternative Technologies

Government intervention is justified when private enterprise shows a strong tendency to be limited to production and distribution of products that embody a high degree of monopoly, such as chemicals or hybrid seeds. In these cases governments may legitimately support the introduction of alternative technologies, such as integrated pest management (IPM), and encourage producers of truthfully-labelled seeds of composite varieties, and of disease-free planting material of root and tuber crops, which can be easily reproduced by small scale farmers.

Spreading the know-how on IPM may be a publicly-supported research and extension activity, justified on grounds of mitigating negative externalities and of enhancing competition in the plant protection technology market. Supporting producers of composite seeds and of disease-free planting material fills a gap in supplies and stimulates competition with hybrid seed suppliers. It also responds to the needs of a different category of users.

The most effective way of providing government support depends on country- and location-specific conditions. However, in the majority of cases, some form of devolution to the private and voluntary sectors would be useful, with the public sector retaining such functions as providing technical advice, supply of basic seeds, supervision of truthful labelling by producers, and implementing phytosanitary controls at government research laboratories.

Local governments do not appear to be an adequate locus for decision making in matters related to input supply and crop marketing services, particularly with respect to subsidy management, because of spillover effects. For example, products subsidised in an area can be sold at market price in another area.

A strong case can be made for central governments to support capacity building and training of CSOs in handling input supply and crop marketing services. Apex organisations in the voluntary sector can play an important role in moderating the impact of asymmetric information, which affects farmers and particularly poor farmers in a negative way with respect to both input purchasing and crop sales.

A most important contribution of government to mitigate rural market failures is to improve transport infrastructure. Many development specialists would share the view that if all the financial and human resources spent on fostering government direct intervention in crop marketing, input supply, and related credit operations were spent on improving rural transport infrastructure and on designing effective mechanisms for the maintenance of secondary and feeder roads, then the history of agricultural and rural development might have recorded a different performance in developing countries.

Provisions of Transport Infrasructure

Decentralisation may provide important potential solutions to the provision of transport infrastructure. In most countries road construction and maintenance is the responsibility of public works departments (PWD). Normally, however, the PWD is not responsible for rural roads, which creates a serious institutional gap through the absence of a clear allocation of authority and resources in this domain. Planning trunk roads by PWD is generally influenced by the importance of urban centres, with little account being taken of the agricultural potential of rural areas.

Traffic counts are used to establish priorities and these record existing situations. This invariably means that high potential agricultural areas with inadequate transport infrastructure produce much less, and therefore generate much less traffic, than they would with better infrastructure. This is well known to agricultural economists and planners and there is ample empirical evidence of the impact of rural road programmes on crop and livestock production. However, strong arguments against giving priority to feeder road construction are the difficult logistics and the high maintenance costs, particularly in tropical and equatorial environments.

Devolution of responsibility for planning, construction and maintenance of rural roads is one option tried in several countries with World Bank support. In some cases, local governments have been devolved financial resources on a matching grant basis for rural road construction and maintenance. These have been financed from a share of the central government fiscal revenue from road users coupled with the authority to raise local levies from road users. The local governments then contract out design, construction and maintenance to various private enterprises. The deconcentrated units of technical ministries provide the necessary support to the local governments in contracting and supervision of consultants, and/or contractors' performance. The advantages of this option are to:

— Provide a stable source of funding to local governments

— Stimulate their capacity to raise local taxes and, presumably, the willingness to pay taxes which taxpayers know are earmarked to provide better services to them

— Entrust planning to the level that is best aware of requirements and potential benefits and to promote local civil work enterprises in the private sector.

There are however drawbacks to this approach that need to be considered and, if possible, overcome. These include:

— Limited technical capacity of local governments

— Inadequate resources or simply resistance of deconcentrated units of technical ministries to provide the required assistance

— Interference by local pressure groups with priority setting

— Low revenue collection from road users' levies.

In practice, it will take time to make the new system functional and performance is unlikely to be the same in all local governments.

A complementary option is to provide funds to finance, through CSOs, local community projects on a matching grant basis under partnership arrangements. This would be more appropriate when road projects benefit single villages rather than a number of villages, which is the domain of the district on account of jurisdictional spillover.

Such small projects could be supported on a cost-sharing basis with the local community's share, in cash or labour, paid in advance to provide proof of stakeholders' commitment. The argument in favour of this option is that if local people are free to select the project they desire they will select a road project whose benefits are not shared with other communities. Having willingly invested their own resources in its construction there is a higher probability that they will take care of the road maintenance.

Support to Communications

Another important contribution of government to improve the performance of private markets is the provision of better communication systems. Mass communication media are increasingly used by public extension organisations. Technology breakthroughs in telephone communications mean, for example, that small portable cellular phones fuelled by batteries can be purchased and operated by farmers at reasonable costs. This would allow them to acquire very valuable information on crop prices, market demands, etc., on a private basis thus reducing the disadvantages of asymmetric information. However, for this to be possible, the networks of such services need to cover rural areas.

Price Stabilisation Schemes

An important aspect of crop marketing that also affects input markets concerns price fluctuations, which are a major problem for agricultural producers. At the global level, commodity markets have developed very sophisticated institutions with a view to attracting private investors and moderating the negative impact of sharp and erratic fluctuations in the price of agricultural commodities.

Examples are global commodity information services, global commodity exchanges, financial products such as trading in futures and ample commercial storage facilities. The overall impact has been positive, but price fluctuations continue to be a feature of the international agricultural commodity markets. The experience of most developing countries with price stabilisation schemes and buffer stocks have been fairly negative, with the result that policy reform advocates have argued convincingly in favour of ending such schemes.

The current opinion is that government intervention in export crops markets is not advisable because little can be done by a national government to influence global markets beyond what global markets institutions already do and the risk of wasteful use of limited financial resource is considerable.

In domestic markets, crop price fluctuations are more pronounced with sharp seasonal fluctuations in food crop prices compounding the impact of inter-annual fluctuations. Price instability is one of the main risks faced by agricultural producers, and the absence of adequate market infrastructure and institutions, particularly for food crops, is a major market failure in developing countries.

Furthermore, failure to maintain national food security is often a threat to political stability. The issue confronting decentralisation reform designers is how to introduce effective mechanisms aimed at moderating agricultural price fluctuations, with a minimum of direct government involvement. There are some examples of successful management of floor price policy by government agents. BULOG, an Indonesian government organisation, controls minimum paddy prices through open market operations involving less than 3 percent of the national crop. It does this through good management of a deconcentrated monitoring of local markets, and of an effective network of purchasing agents, coupled with adequate resources.

Admittedly, paddy is an irrigated crop, with smaller production fluctuations than rainfed crops, which makes price stabilisation easier. Nevertheless the BULOG experience is worth keeping in mind. However, grain stock management and price stabilisation, at any level, require high management skills, transparent and accountable management, no corruption, and good experience of markets reactions and behaviour. Moreover, no such scheme can effectively deal with the impact of long lasting droughts or exceptional calamities.

Decentralisation policies include other successful experiences. These concern, for example, the promotion of community storage schemes at village level in several African countries, some of which have managed to effectively improve producers' prices. Overall, however, the experience of these projects is mixed. Some have been wrongly designed, with an emphasis on physical storage facility construction rather than working capital requirements, and without a workable link to the sources of commercial finance.

Many have been located in deficit areas for food security purposes, rather than marketing improvement. The approach has a better chance of working in grain surplus areas where farmers' stocks of grains can fetch higher prices during the lean season. Group storage has considerable advantages, including the possibility of obtaining credit from commercial banks on the basis of 'bonded stock' procedures, improved market information, and lower transport cost through bulk conveyance to markets. Coordination with the central government stock management policy is required, to avoid decisions about selling public stocks at the wrong time undermining the viability of village storage operations.

There is little justification for government to subsidise inputs and equipment. It is sometimes alleged that there is 'inadequate' use but this is often because technical 'requirements' are confused with effective demand. There are several instances where special crop production programmes associated with input subsidy schemes have performed well, but whether this was primarily determined by input subsidies or by favourable market conditions, adequate infrastructure, and appropriate technologies is far from clear.

On the other hand, there are many examples of overestimation of the potential demand, because many important factors determining demand are invariably underestimated by agricultural development planners. Among these are:

- — High risk factors of various nature confronting producers
- — Poor crop responses to inputs in specific local agronomic conditions
- — The limited share of the cash economy in the farmer household budgets
- — Better alternative investment opportunities for employment of surplus cash resources.

Regulatory Issues

Even if the government divests itself of its responsibilities for producing, delivering and financing input and crop-marketing services, it still has an important role in providing a regulatory framework with rules that govern the trade in agricultural inputs, equipment, and crops. This framework should foster the effective functioning of markets and moderate the effect of market failures. The revision of this regulatory framework is an important part of the decentralisation process.

In a decentralised system of governance, and in pursuit of liberal economic policies, regulations governing trade play a vital role in creating a satisfactory enabling framework. With regard to input and equipment supply and crop marketing services the main issues that need to be covered include:

— Ensuring a competitive environment in the market for a specific good or service
— Ensuring that users of goods and services are informed about the quality of products they buy
— Certifying the quality of the products or the competence of the suppliers where quality is not easily discernible by visual inspection
— Preventing or regulating the use and sale of products that may cause negative externalities.

Trade regulations may be handled at different levels of governance. Typically the central government establishes the regulations or at least the framework within which detailed regulations must be formulated by other agents, and may delegate the power of taking action for enforcement to lower levels of governance. In some cases the authority to formulate regulations may be devolved to lower levels.

Negative Externalities

Trade regulations are particularly important in agricultural input supply to protect farmers from frauds by unscrupulous suppliers, and to prevent or regulate negative externalities. Enacting adequate rules is not a simple matter when complex technical issues are involved. Enforcement is not always easy. Good simple regulations vigorously enforced can be highly beneficial. Decentralised enforcement, and some degree of devolution to professional and consumers' associations, may help making rules more effective. However, too many rules, or vague, confused often contradictory rules may have devastating effects.

Over-regulation may obtain opposite results to the declared objectives of the regulating authority. Typically the scope for decentralisation decreases the more numerous are the regulations, and the more complex the procedures which must be applied. At the same time the scope for abuse by public administration officers increases.

Three particular issues related to trade regulation deserve some more attention:

- — The procedures required to obtain permits and trading licences
- — The prevention of negative externalities
- — The protection of users' rights.

Trade Licences

Trade licences are enforced in every country across the world. Good governance requires that the procedures to obtain a licence are simple; transparent and equitable. However, often the criteria for obtaining a licence are opaque and there are unreasonable delays in granting a licence. In agriculture, the fees charged and the documentation required by the administration to issue a licence can be serious handicaps for small traders. The principle of subsidiarity suggests that responsibility for issuing licences can usually be undertaken by deconcentrated local units of the central administration, or devolved to local governments, with a view to cutting 'red tape' and delays in processing the applications.

Administrative Reforms

Market liberalisation and privatisation in many countries have not always been accompanied by the reform of administrative procedures. Even when far reaching changes in policy and procedures have been introduced, widespread and transparent information has not always followed. The failure to simplify procedures and/or adequately inform people about changes does not reduce the delays, the high transaction costs, and the other impediments which limit the effectiveness of liberal economic policies. As a result, many rent seeking positions in bureaucracies are not seriously challenged and the role of private enterprise will fall short of expectations.

In a democratic system of governance, the public authorities have a precise responsibility to make rules clearly understandable and widely known, well beyond the publication in official journals that are sometimes difficult to find even in capital cities in developing countries. Deconcentration and devolution, when local managers are effectively accountable to a local constituency, may help to reduce the risk to some extent.

Dealing with negative externalities requires that input users are adequately informed about the potential negative effects of the products that they intend to use on the sustainability of their own resource. This is essentially an extension message problem, which, because of conflict of interest, cannot be entrusted to the product suppliers. It also needs users to be fully and very clearly informed about the specific characteristics and the adverse effects of inappropriate use of a product, such as endangering the user's health, or the health of his plants and animals. In part, this can be achieved by rules that insist on adequate labelling of products and on full disclosure of information by suppliers in product marketing. Government has a general policing responsibility in this respect vis-à-vis traders and manufacturers, which needs to be forcefully implemented. The extension service has a specific function of warning uneducated farmers about dangerous products circulating in the market.

Users' Right Protection

This is particularly important in the case of seeds, since product quality is an essential element of the price set by suppliers and a consumer's decision to buy. However, the external characteristics of the product often do not reveal product quality, making fraud easy. Government therefore has the responsibility for regulating the quality of the seeds traded in the market. The rules enacted depend on decisions about the strategies to be followed to enhance quality control, and the related organisational consequences such as delegation to autonomous agents in accordance with the specialisation principle.

Alternative Regulatory Approaches

There is a different view emerging about the role of government in performance and quality control, based on neoclassical economics arguments and on some empirical evidence about the cost and benefits of the regulatory function. Regulations have a direct cost for government and suppliers, that may or may not be commensurate with benefits. Some regulations may yield negative net benefits. For example, delaying the introduction of new products for a long time, until all tests of performance and quality are carried out, deprives users of considerable gains which they may have made by using the product, and discourages private suppliers to seek introducing new products.

Country case study evidence suggests that benefits from regulating input performance or quality are low, while costs are significant. Thus the argument goes 'that the optimal level of this kind of regulation is nil'. However, other benefits of regulation may be significant, such as those of preventing or reducing serious negative externalities. It is therefore suggested that the regulatory function be limited to the latter cases, or to the need to check possible frauds. The market generally would take care of performance and quality by penalising suppliers of inferior products. This policy would prevent the risk that regulators in many transition and developing countries, who distrust private companies and misunderstand the market process, suppress market entry and competition, leaving farmers heavily dependent on public research for new technology and on regulators to judge input quality.

The suitability of this strategy to specific situations, however, depends critically on the stage of market development that has been achieved. It is likely to work mostly with commercial farmers, only for saleable technology products, and in well performing markets. It is debatable whether it would work as well in small fragments markets, where buyers suffer from asymmetric information problems or for non-saleable technology products.

12

Agricultural Technologies and Intellectual Property Rights

The implications of newly emerging technologies for law and legal changes has not been paid as much attention by scholars as it should be, compared to activists who have often stressed on these linkages in terms of their impacts on peoples' abilities to meet their needs and sustain their livelihoods. On the one hand, new agricultural technologies deriving from biotechnology and genetic engineering are often proprietary technologies which cannot by reproduced by users without paying for their use again and again. On the other hand technologies are being evolved which make in impossible for farmers to reproduce seeds even if they wish to.

Both these have impacts on the farming community in terms of reducing their abilities to innovate, experiment, and adapt to ecological and economic changes. This directly reduces their capabilites and impacts on their welfare. Also through spread of these technologies, there is the possibility of genetic contamination, and biodiversity degradation which reduces peoples' abilities to cope and eke out a livelihood in various ways. But the development of technologies also points to a radical new rupture as far as rights issues are concerned. The meaning of legal rights in relation to peoples' capacity to improve their welfare undergoes in transformation along with changes in technological options. In the case of traditional technological regimes functioning under customary and locally negotiated norms, technologies are regulated on the basis of a right to livelihood and sustenance from natural resources. On the other hand, new technologies have led to the creation of a new rights regime wherein the use and exploitation of a technology for making profits is given primacy over all other considerations.

Lakshman Yapa in analysing the "genetic transformation of seed into a nonreproducing commodity" refers to the way in which promoters of these technologies have failed to perceive the social relations of production embedded in all technologies, what he calls "seed as nexus of relations". In critiquing the whole epistemology of development that are endorsed by the supporters of new agricultural seed based technologies, he points to relations of dependence and types of 'scarcities' that emerge as a result of adoption of such technologies.

An analysis of studies of agricultural technologies based on modern biotechnology and genetic engineering reveals some basic differences in the way in which these are critiqued by scholars and activists in the developed and developing countries. Ethical - as differentiated from moral - reasons and environmental and health effects are most often cited by western opponents of these technologies for imposing restrictions on them. On the other hand, activists in developing countries, particularly in India have cited the direct effects on poverty, and indirect effects, in terms of the impacts of environmental degradation on livelihoods, as reasons for opposing these technologies. In particular effects for specific groups such as women, agricultural labourers, and those who are more dependent on natural resources, common property, and diversified farming and livelihood systems are projected as those most likely to be affected. However while activists have quoted selected studies and anecdotal evidence in support of their positions, rigorous studies, similar to those in the west are absent for most developing countries. This is partly due to the methodological complexity of assessing actual and potential long term environmental impacts on peoples livelihoods.

One way in which these methodological complexities could be addressed is to study the impacts in terms of their consequences for peoples sources of livelihoods, the mechanisms by which rural people are able to meet their subsistence needs. We also need to study the capabilities and entitlements that people have or lose as a result of a particular configuration of natural resource management, availability, and technology use. Technologies, through an adverse impact on biodiversity or on commons, may not only reduce livelihood options but also take away the opportunity to use one's capabilities for productive purposes, or take away certain entitlements, for instance the entitlement to exchange one's labour for a set of commodities.

While Amartya Sen's work on capabilities and entitlements are now well accepted in famine analysis, and analysis of food security, few attempts have been made to widen the scope for applying these concepts in the broad areas of development economics, and poverty analysis. Changes in property rights regimes for instance, and shifts in natural resource use due to economic, social, political, technological and environmental reasons offer ample scope for studying their impacts on peoples' welfare. The opportunities for clarity in understanding dynamic processes are all the more since Sen specifically focuses on legal and rights issues in his presentation of the entitlements approach.

Sen's theory of entitlements based as it is on a set of rights of 'ownership, transfer and rectification', brings in law as an important mediating agency in a very significant way. In this framework, 'endowment vectors' are related to sets of alternative commodity entitlements through what is called an 'entitlement exchange mapping'. The entitlement approach centres on an individual's entitlements to commodity bundles that may also include food.

Exchange can be of two types: trade - which involves exchange with others, and production - which involves exchange with nature. Entitlement then depends not only on the endowment vector that an individual starts with, but is also related to exchange relations. These relations in the form of entitlement exchange mapping depend on the legal, political, economic, and social characteristics of the society in question. Entitlements therefore refer to "the set of all the alternative bundles of commodities that he can acquire in exchange for what he owns" which is the "'exchange entitlement' of what he owns".

Sen further states that '"production opportunities, trade possibilities, legal rights to the produce, and social conventions" all affect the e-mapping. While endowments can decline, entitlements can also fail if for instance food entitlement declines because one has produced less food, or one cannot obtain adequate food through trade. Poverty and starvation therefore result from both a "fall in the endowment bundle and unfavourable shifts in exchange mapping". Analysis of legal and technological shifts with reference to natural resources ownership and control precisely enable us to understand how for instance, the decline in access to, and the degradation of common property and biological resources can result in change in ownership bundles, and consequent entitlement failures

An understanding of entitlements would not be complete without the complementary concept of capabilities. For Sen, a capability is a feature of a person in relation to goods. In that sense it is much more than simply 'endowments'. Capability is the ability to function and "reflects what a person can do" In Sen's words, "capabilities are ... directly valuable in a way that the possession of primary goods1 cannot be, since they evidently are means to some more human ends". A capability set refers to the alternatives sets of functioning that an individual has access to based on endowments but also on political, legal, social, and economic structures, and includes such features as freedom, health and education. Sen thus displays explicit awareness of the role of laws and legality both in bestowing and transferring endowments, and in providing an individual with entitlements to meet her / his needs.

Over the last couple of decades, scholars and activists have brought out the implications of shifts in property regimes for peoples' livelihoods and well being. However the exact process by which livelihood sources and abilities to cope are compromised by these changes are not well understood. Moreover the implications of legal and technological changes which undermine pluralistic and diverse approaches and thereby the entitlements and capabilities of individuals, households and communities have not been adequately worked out. By focusing on the shifts in entitlements and capabilities arising from legal and technological changes, this chapter attempts to explain how people's abilities to cope, experiment, innovate, adapt, and manage natural resources as part of survival strategies have been deeply impaired.

Agricultural Technologies and Intellectual Property Rights

Developments in biotechnology and genetic engineering have introduced new options in agriculture. Many of these have controversial effects, and are being opposed by environmental activists around the world. However while the opposition to these new technologies in agriculture have come from activists, scholarly studies charting the actual impacts for peoples' livelihoods have been few partly because of the low levels of adoption of these technologies. In this section we attempt to logically develop some of the consequences of these technologies for peoples' capabilities and entitlements, and secondly, to discuss the ramifications of these for intellectual property rights.

New IPR regimes threaten people's abilities to cope, adjust and innovate in order to adapt to changes. Environmental consequences of these new options such as loss of biodiversity also significantly impair community resilience in the face of stresses and shocks, and environmental, demographic and economic changes.

Modern technologies, despite their benefits, perpetuate inequalities among groups within a community and between nations and economies. This occurs through excluding people from access to forms of knowledge, skills, techniques, and markets, which are important for subsistence, survival and for competing in a globalised economy. Despite, or precisely because of large-scale adoption of new technologies, yields are rapidly declining in most parts of the world in agriculture, which is compounded by loss of soil fertility, and increased pest attacks. The impacts of these are most dramatically seen in the suicides of farmers in parts of India, over the past few years, attributed to indebtedness, loss of control over the production process, and the unsustainability of their livelihood patterns. However the focus here, is specifically on the products of modern biotechnology and genetic engineering including GMO (genetically modified organisms).

Several environmental consequences can occur as a result of using GMO seeds. For instance, the influence of a genetically engineered organism on the food chain may damage the local ecology. The new organism may compete successfully with wild relatives, causing unforeseen changes in the environment. Once genetically modified bacteria and viruses are released into the environment it is impossible to contain or recall them. Unlike chemical or nuclear contamination, negative effects are irreversible. Toxins introduced into plants to resist or kill pests may end up killing beneficial species such as butterflies, bees and beetles as well. Such crops have the capability to accelerate evolution of resistance in pests and the evolution of 'super pests'. Once pests become resistant to pesticides, it may lead to higher use of pesticides and higher rates of crop failure.

This will not only result in more debts and more suicides but will wipe out both biodiversity and farmers' livelihoods. Some scientists however state that plants genetically engineered to be herbicide-resistant will greatly decrease the amount of herbicide use. However, farmers, knowing that their crops can tolerate the herbicides may use them more liberally, leading to extermination of beneficial plants and organisms. By increasing production costs through increased use of purchased GMO seeds and inputs, these technologies thus may have deleterious effects on the livelihoods of farmers in developing countries. They also increase dependence on external agents such as seed and pesticide dealers.

Insects, birds, and wind can carry genetically altered seeds into neighboring fields and beyond. Pollen from transgenic plants can cross-pollinate with genetically natural crops and wild relatives. All crops, organic and non-organic, are vulnerable to contamination from crosspollination. In the case of genetically engineered seeds, all the seeds have identical genetic structure.

As a result, if they are planted over a wide area, and if a fungus, a virus, or a pest develops which can attack this particular crop, there could be widespread crop failure. Not only do these problems raise the issue of genetic pollution, but also as we shall see, fundamental issues relating to protection of indigenous knowledge and products vis a vis intellectual property rights of scientists and corporations arise. The existing means of resolution of such conflicts are invariably biased towards MNCs.

Apart from the scientific / technical differences which have consequences for the environment, plant behaviour and human metabolism, other differences emerge when the principles of modern biotechnology are put into practice. It is interesting to note that differences in social and economic impacts noted by scientists in third world situations due to modern technologies, are not even mentioned by those in the first world. The implications for developing countries include those for biodiversity, problems related to technology transfer, knowledge and skills retention for women, and the role of privatisation in influencing the choice of research area in terms of commercial versus social considerations.

Arguments for 'substantial equivalence' then, are derived from a narrow, laboratory based, reductionist view of the science of genetic engineering. Laws and regulations relating to the testing, use, production and patents of these technologies then do not even consider the social and economic consequences of these technologies, and are only concerned with certain kinds of private property rights. Thus assessments of costs and benefits of these technologies do not touch upon many social, economic, and cultural issues of communities which are affected by the new technologies.

While environmental consequences of technologies inevitably have impacts for communities dependent on them, the situation is starker in developing economies where there is a greater degree of dependence on natural resources for survival and subsistence. For instance the loss of biodiversity has an impact in terms of increased abiotic stress. Farmers may have a reduced choice of mechanisms in terms of varieties and options for adapting to drought, low fertility soils etc. Under conditions where much of the crop originates from seed bought from companies, seed renewal is not adequate, since the quantum of saved seed is less, and hence there is a deterioration in the quality of seeds. Gradually farmers are forced to depend more and more on biotech companies for their seeds. In many parts of the world there are already reports that farmers plant GMO's regardless of market demand because there are not enough regular seed to go around.

Such large-scale homogenisation resulting from increasing adoption of genetically engineered and transgenic crops are likely to worsen the ecological problems already associated with monoculture. Unquestioned expansion of this technology into developing countries may undermine the agricultural diversity of many of these countries, and extensive monoculture could inhibit or reduce this diversity, especially keeping in mind consequences in terms of serious social and environmental problems.

Monopoly capital in the agri-food business has a tendency to create broad international markets for single products, thereby simplifying cropping systems and creating genetic uniformity in rural areas. Large areas planted to a single crop variety become quite vulnerable to new matching strains of pathogens or insect pests. Moreover, widespread use of homogeneous transgenic varieties will inevitably lead to "genetic erosion," and a sharp reduction in biodiversity.

Much of modern technology ostensibly at the service of the poor then, even if not by design, at least by ignorance, have reverted back to the much reviled and derided 'top down' approach of development. Small and marginal farmers typically plant several different varieties on their land, tailoring their choice to the characteristics of each plot, in terms of drainage, fertility, consumption needs etc. Planting many varieties is also a technique of risk minimisation for small farmers.

However, such options cannot be easily developed with current top down and highly centralised research and extension structures and methods. Formal research methods and research systems simply are not equipped to handle the complexity of physical and socio-economic conditions in much of third world agriculture. Thus numerous variables important to farmers have to be 'reduced' or ignored altogether in order to produce new technologies. In particular, studies have shown that small farms utilise "a broad array of resources ... and their farming systems are diverse" focusing on preservation of biodiversity and ecosystems.

By contrast, biotechnology and genetic engineering focus only on single traits: and ignore multiple livelihood systems. The technolgoies developed therefore may not provide the necessary inputs or resources, such as fodder, food, fuelwood, building material etc. necessary to support small farmers' livelihoods.

Genetic engineering is the very antithesis of recent approaches such as agroecology, farming systems or participatory, farmer-led research. A purely technological approach ignores real alternatives. It ignores the mixed crop livestock systems, and the use of indigenous seeds and inputs. HYV and GE are approaches which are reductionist; and are decontextualised from the specific systems in which farmers eke out their livelihoods. At best improvements are at the cost of other components of a livelihood or farming system. At worse, it exacerbates existing problems.

iversity was and is a central feature of the old strategies whose function was to share scarce resources such as water, nitrogen, soil nutrients, light, and energy resources. In the specific context of developing countries, not only do their economies become more and more dependent on the metropolitan economies, leading to further underdevelopment, but significant sections of society get marginalised and excluded from the production process which adversely affect their well being, and disempowers them by taking away their capabilities to function and earn a living for themselves.

Historically in order to survive under such conditions, farmers have attempted to adapt agricultural technologies to their highly variable but singular circumstances, in terms of agroecological, socio-economic, and environmental factors including resource bases, access to institutions, and climatic conditions. Farmers have thus evolved complex sustainable livelihood systems which balance risks related to drought, market failure, pests, etc. with factors such as labor, investment, nutrition, weather patterns, etc. The agrarian systems that evolved were characterised by multiple annual and perennial crops, animals, fodder, and a variety of foraged wild products.

Under such highly varied circumstances, uniform varieties, or precision techniques such as those introduced under the green revolution, and genetically engineered or 'transgenic' innovations, are unlikely to be useful for many such farmers. On the contrary further marginalisation of the rural poor is likely. Studies carried out in arid Western Rajasthan, in India for instance confirmed that villages with low genetic diversity were unable to cope in conditions of drought, compared to village where farmers had saved seeds of traditional varieties of pearl millet better adapted to weather conditions. Low genetic diversity in the former villages had been brought on by high adoption of hybrid cultivars derived from sa limited gene pool, and an abandonment of traditional or local landraces.

The social and economic consequences of loss of biodiversity are quite severe. Farmers are unable to innovate and adapt to changing conditions, drastically bringing down their capability levels. The endowments that they do have begin to yield less, leading to both direct and trade entitlement failures. Farmers are forced to be part of globalisation since industrial agriculture, export orientation and commercialisation lead to homogenisation and a preference for standardised products. The increasing difficulties of transnational agribiotech firms in maximising their profits compounded with huge amounts spent on research for which there are little returns, has meant that they develop varieties with wide adaptation. This also results in loss of biodiversity Farmers are forced to be part of this process of reducing biodiversity through market mechanisms of a decline in prices for local varieties, and pressures to adopt new technologies for their potential gains.

Several of the innovations of biotechnology directly threaten the independence of farmers around the world. Vasavi, commenting on cotton farmers suicides in southern India indicates that commercialisation of agriculture and/or the introduction of technological innovations result in the 'separation of the economic dimension of local agriculture from its established cultural bases,' since the social implications in terms of 'disjunctions' and 'dissonances' within society", are not considered. In "privileging the economic impetus of taking to commercial agriculture, many had overlooked the importance of retaining social ties", she argues pointing to decrease in social capital within the village and increasing dependence on external commercial agents.

Since biotech companies have invested millions of dollars in research, they naturally wish to get back returns at the earliest. When farmers buy seed of genetically engineered varieties, they have to sign a contract that outlines the rules of use. For example, they are not allowed to save any of the seed from their crop to plant the next season. Saving seed is a common practice used for thousands of years, but companies like Monsanto forbid saving, reselling, or exchanging seed, requiring farmers to buy new seed from the company every year. Companies can legally patent biotech crops because the plants are basically creations of the company-there is no way they could possibly exist naturally. They contain genes from bacteria or other organisms spliced together in laboratories.

Monsanto for instance has taken several farmers to court over this issue and has accused farmers using saved seeds in Canada and the US of infringing on their intellectual property rights, but many of the farmers claim the wind blew the GMO technology into their fields which is difficult to disprove. Thus the sustainable practice of saving seed and cultivating the land now has become an unsustainable form of livelihood due to the forced dependence on an external agent from whom inputs are to be purchased. Farmers then cease to have complete control over the production process, and also cease to have the ability and freedom to experiment, since that process can only be carried out by the owner of the technology.

Farmers and farmer organisations in India have also expressed fears about the enslavement of farmers to multinational and local corporations and a loss of freedom and choice in their cropping patterns and sources of livelihood. The family farms of the poorer nations depend on saved seed for survival. They are too poor to buy new seed every year. Biotech companies also patent technologies and products in the development of which they have no hand. American corporations have patented Basmati rice, neem, and quinoa, taking advantage of indigenous knowledge and centuries of selective breeding by small farmers without giving anything in return. Thus there are very real fears of farmers losing access to traditional knowledge and resources and becoming enslaved to MNC who patent varieties in which they have very little contribution. The loss of knowledge - which is a key community endowment and is used to prevent entitlement failures by providing the ability to adapt, is major outcome of modern technologies and laws (in the form of patents and IPRs) associated with them.

New international and national level laws and agreements are supplemented by technological tools to get greater control over agricultural production and prevent local level adaptations and innovations. Contracts are very difficult to enforce in developing countries and also involve costly litigation. Transnational biotech firms therefore have come up with new sets of technologies which make IPRs redundant. These are usually referred to as Genetic Use Restriction Technologies. This is most clearly reflected in the Terminator Technology, which causes crop seed to become sterile at harvest time. A terminator seed will grow, but the seeds it produces are sterile, thus effectively killing the process that let farmers save and sow their own seeds. This will entail disempowerment of farmers, increased dependence on corporations, loss of traditional skills and indigenous knowledge, and a decrease in the ability to experiment and innovate.

In addition to the above processes, changes have been brought about in most countries with reference to PBRs or Plant Breeders Rights. Farmers rights in most laws and conventions are fragile. Earlier in most of these act farmers rights and privileges in saving, storing and reusing seed were granted. These rights have gradually been eroded through many national and international treaties and acts. For instance farmers' rights in the Plant Variety Protection and Farmers' Rights Act in India has been diluted and they cannot now sell seed. This severely curtails farmers' abilities to experiment and innovate especially in conditions when new technologies fail and farmers have few other alternatives. Many studies have demonstrated the importance of local seed systems in sustaining agricultural systems in the absence of well developed public and private sector seed systems. Thus the consequences of the ban on farmer to farmer sale of seeds effectively means that even when appropriate technologies and options are available, households and communities will not be able to use their endowments to exchange them for appropriate commodity bundles, due to lack of access to these options such as seeds.

In addition to the Terminator Technology, there are a whole series of technologies that are labelled by critics as Traitor Technology. The thrust of these technologies is that for the plant to grow and yield up to it's potential, farmers have to apply various company supplied sprays in conjunction with various company herbicides/ insecticides to turn on genetic 'switches' that control specific characteristics such as high yield or pest resistance. Most innovations in agricultural biotechnology are profit-driven rather than need-driven. This is best illustrated through a review of the major technologies on the market today such as herbicide resistant crops like Monsanto's "Roundup Ready"soybeans, which are tolerant to Monsanto's own herbicide Roundup, and Bt" crops which are engineered to produce their own insecticide.

In the first case, the objective is to win a greater herbicide market-share for a proprietary product. In the second it is to boost seed sales at the cost of damaging the usefulness of a key pest management product - the Bacillus thuringiensis based microbial insecticide which is used by many farmers, including organic farmers, as a powerful alternative to insecticides. These technologies have been developed by biotechnology companies to intensify farmers' dependence upon seeds protected by intellectual property rights, which conflict with the traditional rights of farmers to reproduce, share /sell or store seeds.

Thus corporations will force farmers to buy their brand of inputs and will forbid farmers from keeping or selling seed. By controlling germplasm from seed to sale, and by forcing farmers to pay inflated prices for seed-chemical packages, companies are determined to extract the most profit from their investment. Laws now become infructuous as technologies take over their function. A key threat to legal pluralism then emerges from modern technologies.

Another technique to control farming operations is by linking chemicals and seed development, and by developing technologies to protect seed and plant from their own chemicals, corporations seek to accelerate increases in expenditures per unit for seeds plus chemicals, delivering significantly lower returns to growers. Companies which are developing herbicide tolerant crops are trying to shift as much per unit cost as possible from the herbicide onto the seed via seed costs and/or technology charges. Increasingly price reductions for herbicides will be limited to growers purchasing technology packages. For multinationals it is infinitely more profitable to sell seeds programmed to commit suicide at harvest. This will force farmers to pay the company in order to obtain the chemicals to have them re-activated for the next planting either through a seed conditioning process or through the purchase of a specialised chemical that brings saved seed back to life.

Such technologies also shift the cost of developing and producing seeds to the farmer. The seed companies will only have to sell seeds and not produce, transport, or stock them. Farmers will reproduce seeds, and either through royalty / license fee mechanism or through chemically induced germination, the seeds will be made 'fertile'. As these seed companies increase their control of the world market, there will be diminished interest in future plant breeding and research. Furthermore, farmers will not have any power over what to grow or plant and will be "in a position of absolute dependency" on multinational seed companies. In line with industrial workers in developing countries producing products for the global economy who have become part of the putting out system, farmers also will become putting out workers with little control over their skills, tools, etc. Such technologies thus directly reduce capabilities through a deskilling process whereby gradually, farmers - especially women who are mainly involved in seed selection and saving - lose the ability and freedom to innovate.

Knowledge of germination requirements, seed preparation, weather, soil, breeding of farm animals, knowledge of the feed value of fodder species, fuel of fire wood types all have vested with women traditionally. Enormous disempowerment and exclusion from the economy and decision making thus occur for women due to the introduction of certain kinds of new technologies. Thus genetically engineered seeds seem to be designed for agribusiness farming, not for the capabilities of small family farms of the developing nations like India. How are they to buy and distribute the required chemical inputs? What will happen to their livelihoods if they are made dependent for all their inputs on multinational firms? How will farm households manage risk in conditions of uncertainty and loss of diversity - which is a major source of risk management?

Perhaps the most significant marginalisation that can occur as a result of the flow of genetically engineered products is of rural women involved in agricultural operations. Agarwal has argued that "womens' access to land and livelihood ..... "is also important for improving productive efficiency" on farms. Loss of biodiversity, influx of new options accompanied by a decline in the overall choices available are most likely to affect women living and working on farms. Two broad impacts on women can be identified. At a first level, traditionally in most Asian countries, women have been the repositories of knowledge, skills and techniques in agriculture, especially those pertaining to saving and storing seed, but also, and more importantly, breeding for specific traits.

The 'hybrid' revolution forcing farmers to buy seed has already severely impacted on these skills and knowledge base, which has been eroded. The complete appropriation and alienation of the capacity to innovate and experiment (en)gendered in genetic engineering and biotechnology will further exterminate these skills and forms of knowledge. At another level women also will suffer from loss of biodiversity and an increase in monoculture, especially a shift away from mixed cropping and crop-livestock systems. This will accelerate ecological degradation thereby making it difficult to obtain water, fuelwood, and fodder - all activities carried out by women. This impacts on time-allocation patterns with consequences for larger development issues including health, employment, child-care, education and so on.

The ability of women to better manage households under conditions of scarcity is also compromised as sources of income in cash and kind dry up. Traditional cropping systems catered to a variety of needs - food, fodder, and fuelwood, which is not the case with monoculture, and with varieties bred for specific traits which normally do not meet other needs. Large scale decline in the livestock economy in India, especially the absolute numbers of cattle, are directly related to decline in fodder availability resulting from shifts in cropping systems. It is to be noted that in traditional mixed crop - livestock systems, output and income deriving from livestock vested primarily with women.

Entire communities and nations are thus pushed to the margins of survival, because of reductionist perspectives which ignore the complex survival and livelihood systems of the poor in developing countries. An example which clearly brings out the implications of loss of biodiversity for reduction in capabilities in a major way is the recent controversy over 'golden rice'. Iron, protein and vitamin A deficiencies are the chief causes of malnutrition in much of Africa and Asia. These result in anemia, impaired learning ability, increased susceptibility to infection and reduced capacity to work. The consequences also include reduced life spans, and disabilities such as blindness. Recent innovations in genetic engineering include rice varieties which are fortified with iron or vitamin A and which have been touted as significant achievements for modern biotechnology. One of these is 'golden rice' which is a genetically engineered rice variety fortified with Vitamin A.

Nutritionists argue that vitamin A deficiency warns us of broader dietary inadequacies associated with poverty, as well as with agricultural change toward monoculture. People's diets have been reduced to rice because of this and thus people suffer many dietary illnesses. It has been observed that Vitamin A deficiency is more widespread in those parts where rice or wheat forms an integral part of the diet compared to areas where the major staple is maize, millet or sorghum. These in fact provide considerable amounts of vitamin A and other nutrients including higher levels of protein.

Unimaginative poverty reduction policies such as the Public Distribution System in India have aggravated this problem by replacing traditional cereals and providing the poor with rice or wheat at cheap prices. Thus growing poverty has resulted in the poor becoming more dependent on rice alone for their dietary requirements. The problem therefore isn't one of rice not containing a particular nutrient, but the absence of dietary diversity resulting from poverty and monoculture.

The issue that is highlighted here is that in areas where previous green revolutions have taken place, reduced dietary diversity is an outcome of reducing cropping diversity. Also in the green revolution areas, skewed use of chemical fertilisers especially focussing on yield enhancing fertilisers such as nitrogen, has meant a concomitant neglect of micronutrients. These have had their effects on food quality, diet and nutrition. Reduced capabilities in terms of health are therefore related to loss of biodiversity. While balanced nutrition is more and more becoming a problem in rural areas, loss of biodiversity also has reduced access to traditional plants which were used for medicinal purposes.

There are many other ways in which entitlements of individuals and families are destroyed through newly developing agricultural technologies. In particular there are significant implications for what Sen refers to as a decline in "trade independent security" A study in Andhra Pradesh in India has indicated a rapid increase in female child labour in areas where seed production of high technology cotton seeds are produced. Women and female children have been taken out of family farms producing food crops to participate in cotton seed production. Other scientists have also expressed concern that farmers may lose their independence and become agricultural workers on contract to multinational agribusiness corporations. Thus even though they may have access to incomes, in times of scarcity, distress or shortage, they have entitlements deriving from direct productive activity. As Sen has mentioned landless labourers are most likely to face entitlement failures leading to starvation.

Another important issue that is well summarised by the political scientist Arun Agarwal is that decline and devaluation of traditional indigenous knowledge also leads to the devaluation of traditional authority and institutions with implications for conflict resolution.

Power equations change as those with access to 'modern' knowledge have the ability to dictate terms. In a Foucauldian world where knowledge is closely enmeshed with power, loss of knowledge / power can mean significant reduction in entitlements in terms of production declines, bargaining capacities, and cooperative behaviour in the realm of production and exchange.

Rice et all also speak of a 'social infrastructure' that 'shapes seed and information flows', which is likely to be destroyed when seed production is shifted away from farms to firms. Thus while some scholars gloss over capabilities effects by stating that GMO technology "substitutes for human capital at the farm level", the actual effect is one of throwing existing skills into disuse, and reducing the capabilities to adapt, innovate and experiment. In conditions of high variability and risk, this loss cannot simply be substituted by technologies. Nor can such deskilling be compensated by processes such as bio-prospecting contracts. Given that authorship with reference to genetic resources rests in large collectivites, contracts as they are practised benefit only a small number of people excluding large numbers, provoking "egregious and dangerous inequitieswithin regions with genetic resources" as Brush argues.

In fact he goes on to argue against IPRs as they are currently in place, by saying that lack of private ownership enhances adaptive capacities of farmers in coping with risks. One of the important effects of these technologies is that farm level adoption has impacts that go beyond the indivdual farm, impacts occur on local ecosystems, in neighbouring farms etc. In particular, through lack of ability to produce certain livelihood requirements (fodder, fuelwood), decline of ecosystem biodiversity, decline of genetic diversity in a species, increasing tolerance to pesticides among pests etc., it is likely that there is a comprehensive decline and deterioration in quantity and quality of common property resources. These also have substantial implications for social exclusion of people dependent on commons for their livelihoods especially in times of distress and scarcity.

Common Property Resources

Loss of biodiversity can have several kinds of consequences on livelihoods, and the kind of endowments people can exchange to meet their needs. While the loss of biodiversity can affect farm households at the individual farm level, it also has significant effects on common property resources which can impact on basic capabilities and entitlements of individuals. Recent research in this area from around the world, but especially in South Asia have brought out the crucial significance of common property resources for subsistence of the rural poor.

However the actual processes involved in the livelihood strategies of the poor who are dependent on CPRs is still not well known. While direct impacts of decline and degradation of CPRs have been brought out in several studies, what is not so well analysed is the way in which decline of CPRs affect other kinds of entitlements - both direct and trade entitlements. Particularly the links between biodiversity loss and decline of common property resources have not been adequately researched.

The implications for productivity decline, labour entitlement, and declining livelihood options are the focus here, as is an attempt to understand the different legal issues involved. It is important to remember that "there exists a level of complimentarity between common property resources and private property resources" and that decline of CPRs reduce entitlements and capabilities at an individual or household level. Jodha has pointed out that "performance of crop farming is better" with higher access to CPRs; Beck has exposed an important ambiguity in the difference between private property and common property by pointing to the "use of private resources in common" including those that are negotiated by the poor: access to fallen fruits, gleaning of grains from privately owned farm land etc.

To begin with a brief statement on the decline of CPRs - it is estimated that twelve percent of the poor peoples income in India is derived from CPRs which include community pastures, forests, wasteland, ponds, rivers, rivulets, their banks and beds. Jodha, who has written extensively on the decline of CPRs is of the view that this decline not only represents the degradation of a community asset but also leads to an "erosion of survival options".

The landless labourers and marginal farmers benefit most through enhancement of access to CPRs since they need to purchase less of fuel and fodder. In the absence of a capacity to buy, and in a scenario of CPR degradation, the poor, especially women have to travel longer distances and spend more time in searching for fuelwood and fodder. Not only does this decrease the labour entitlement of women, but it also leads to a decline in the amount of time available and allocated for other activities which may be more welfare inducing including wage labour, education etc.

Decline or degradation of CPRs thus takes away the right to ones own labour or reduces its entitlement. Pasha, in studying the use of and access to fodder from common waste land, points out that "animal husbandry provides employment to surplus labour", especially old people and children whose labour would otherwise have gone 'waste. Stating that "animal husbandry is no longer neutral to economic status of individuals" as a result of degradation of village commons, pasture and grazing lands, Karanth also mentions the loss of time for other more productive activities due to increased amount of time spent on foraging for fodder and grazing livestock in distant grazing lands.

The multiple uses of CPRs and the issue of interrelated rights on CPRs is a constant theme in several studies on common property resources. Failure to consider these in new rules, laws and institutional arrangements is frequently cited as a factor adversely affecting those dependent on CPRs for their livelihood. Jodha, for instance points to the possibilities for diversification and flexibility due to enhanced availability and access to CPRs, while also drawing attention to the interlinkages among different production systems - crop, livestock, harvest from CPRs etc. Thus what emerges clearly is that degradation and / or loss of access to the commons does not just result in loss of those resources that people had access to, but also indirectly affects other livelihood systems. They also lead to the inability to use ones own labour and harvest CPR products for exchange and consumption.

What is also emerging in a fairly obvious manner from the studies is that common property is a highly nuanced term and includes resources which are conventionally and by law regarded as private property. For instance, in many parts of the country, farmers, labourers and cattle owners have grazing and animal penning rights on privately owned farmland. Kerr and Sanghi argue for a "need to relax laws restricting harvesting and transporting trees from private land", which was a traditional practice in the semi-arid villages they studied. Reference has already been made earlier to other kinds of common and negotiated access to private property such as gleaning of grains. In conditions of instability in agricultural output for the poor then, CPRs constitute a major source of livelihood. While privately owned resources are sometimes negotiated by the poor for common access, the health of CPRs also affects the efficiency of cropping systems through preventing soil erosion for instance, or by enabling farm households to go in for marketable options since certain needs are met from the commons.

Thus healthy CPRs and access to them enable poor households to make use of their endowments and get fair entitlements by using their resources, and capabilities, including labour, skills, and knowledge in an optimal manner. This requires appropriate legal and institutional support which enjoys local support and legitimacy. Coming to the relationship between biodiversity loss and common property resource degradation, scholars have pointed out that" events or conditions occurring at a particular position in environmental space lead to consequences elsewhere in environmental space."

Bibliography

Bationo, A., C. B. Christianson and M.C. Kalij, "Effect of crop residue and fertilizer use on pearl millet yields in Niger", *Fertilizer Research,* 1993.

Elwell, H.A. and Stocking, M., "Loss of soil nutrients by sheet erosion is a major farming cost", *Zimbabwe Science News,* 1988

Ess, D., & Morgan, M., *The Precision-Farming Guide for Agriculturists,* (1st ed.). Moline, IL, 1997.

Franzluebbers, K.L.R. Hossner and A.S.R. Juo, "Integrated Nutrient Management for Sustained Food Crop Production in Sub-Saharan Africa", *TropSoils/TAMU Tech. Bull No. 98-03,* Department of Soil & Crop Sciences, Texas A&M University, 1998.

Janssen, B.H., "Integrated nutrient management: the use of organic and mineral fertilizers", *In: The Role of Plant Nutrients for Sustainable Crop Production in Sub-Saharan Africa,* H. van Reuler and W.H. Prins (Eds.). Ponsen and Looijen, Wageningen, The Netherlands, 1993.

Juo, A.S.R., and H.C. Ezumah, "Mixed root crop systems in wet sub-Saharan Africa", *In: Field crop Ecosystems,* C.J. Pearson (Ed.), Elsevier, Amsterdam, Netherlands, 1992.

Kang, B.T., "Cropping systems and soil fertility management in the humid and subhumid tropics with special reference to West Africa", In: Management of Nitrogen and Phosphorus Fertilizers in Sub-Saharan Africa, A.U. Mokwunye and P.L.G. Vlek (Eds.), *Proceedings of a Symposium,* Lome, Togo, March 25-28, 1985.

Kang, B.T., and A.S.R. Juo, "Effect of forest clearing on soil chemical properties and crop performance", *In: Land Clearing and Development in the Tropics,* R. Lal, P.A. Sanchez. and R.W. Cummings, Jr. (Eds.). A.A. Ballema, Rotterdam, Netherlands, 1986.

Munyati, M., "Conservation tillage for sustainable crop production: Results from on-station and on-farm research in Natural Region II (1988 – 96)", *The Zimbabwe Science News,* April/June 1997.

Okigbo, B.N., "A review of cropping systems in relation to residue management in the humid tropics of Africa", In: Organic Recycling in Africa, *FAO Soil Bulletin* 53, FAO, Rome, 1980.

Peat, J.E. and K.J. Brown, "The yield responses of rain-grown cotton, at Ukiriguu in the Lake Province of Tangagyika", I The use of organic manure, inorganic fertilizers, and cotton-seed ash. *Empire Jour. of Experimental Agriculture,* 1962.

Ruthenberg, H., *Farming Systems in the Tropics,* Clarence Press, Oxford, UK, 1976.

Taylor-Power, A., "Integrated Management of Agricultural WatershedS: Land Tenure and Indigenous Knowledge of Soil and Crop Management", *TropSoils Bulletin No. 91-04,* Department of Soil and Crop Sciences, Texas A&M University, College Station, Texas, 1991.

Vogel, H., "The need for Integrated weed management systems in small-holder conservation farming, "Rep. No. 11, *Contill project Research Report,* Institute of Agricultural Engineering, Harare, 1994a.

Vogel, H., "Weeds in single-crop conservation farming in Zimbabwe", *Soil and Tillage Research,* January 1994.

Yeung, A. K. W., & Lo, C. P., *Concepts and Techniques of Geographic Information Systems,* Upper Saddle River, NJ, 2002.

Index